English for the

ENERGY INDUSTRIES

Oil, Gas and Petrochemicals

COURSE BOOK

PETER LEVRAI
WITH FIONA MCGARRY

Published by
Garnet Publishing Ltd.
8 Southern Court
South Street
Reading RG1 4QS, UK

www.garneteducation.com

This edition first published 2006
Reprinted 2008, 2009

ISBN 1 85964 911 4

British Library Cataloguing-in-Publication Data
A catalogue record for this book is available from the British Library.

Production

Project manager:	Francesca Pinagli
Editorial team:	Francesca Pinagli, Maggie MacIntyre, Richard Peacock
Design:	Robert Jones
Illustration:	Doug Nash
Photography:	Language Solutions

This course book is dedicated to Averil Bolster, without whom it would never have happened.

Thanks also go to Ian Peart, Mark Trier, Crispin Tucker, Jessica Roberts, Martyn Bingham, Debbie McCormick, Sean Barry, Troy Nahumko, Andy Buckworth, Steve Prior, Nick Hudd, Stan Bennion, Tony Lees, Phil Smith, Abdurrazag Abozaid, Lala Mikailova, Vusale Aliyeva, Evelina Proshina, Narmina Ibragimova, Diane James, Michael O'Sullivan, Tim Bates, Gary Holmes, Irada Suleymanova, Bakhtiyar Mehtiyev, Nadezhda Gulyayeva, Elshad Abdullayev, Ramilla Mammadova, the technical trainers of TTE, and the teaching staff of ELS Baku and ELS Algeria.

BP's Golden Rules of Safety on page 131 are used with the permission of BP plc.

Every effort has been made to trace the copyright holders and we apologize in advance for any unintentional omissions. We will be happy to insert the appropriate acknowledgements in any subsequent editions.

Audio production: Matinée Sound & Vision Ltd.

Printed and bound
in Great Britain by Cambrian Printers Ltd, Aberystwyth, Wales.

Contents

UNIT 1 GIVING BASIC INFORMATION

The aims of this unit are to:

- introduce key terms from the oil industry
- give an introduction to social English – the English you will use to communicate with colleagues

By the end of this unit, you will be able to:

- introduce yourself and ask questions about personal information
- ask the trainer for clarification and repetition
- describe objects, equipment and jobs in the oil industry

Lesson 1: Talking about yourself

A Listen. What is Alan saying? Put the sentences in the right order (A–H).

1 My son, Adam, is 13 and my daughter, Sophie, is 10. ☐
2 They aren't with me in Azerbaijan. ☐
3 Let me tell you about my family. ☐
4 I'm married with two children. ☐
5 I'm a technical trainer. ☐
6 Hi. My name's Alan. A
7 They're both students. ☐
8 My wife's name is Anna. ☐

B Write out the sentences in exercise A in the correct order. Make one paragraph.

____

____

____

____

____

____

____

C Complete the table.

Present simple: *be*	
	I am
you're	you are
	he is
she's	
it's	
we're	we are
	they are

D Write three more sentences about Alan or his family.

1 His wife's name is Anna.

2 ______________________________

3 ______________________________

4 ______________________________

E Talk about yourself. Use these prompts.

(25) years old | married/single | a technical trainee | from (Baku)

F Write three sentences about you.

1 ______________________________

2 ______________________________

3 ______________________________

Lesson 2: Introducing people

A There are five mistakes in the dialogue. Find and correct them.

Ahmed: Hello, Bob.

Bob: Hi, Ahmed. How are you?

Ahmed: Thanks. And you?

Bob: I'm very well, thanks.

Ahmed: You do know Yusef?

Bob: No, I don't.

Ahmed: Bob, this is Yusef. Yusef, here is Bob.

Yusef: Please to meet you.

Bob: Pleased to meet you, too. Do you work with Ahmed?

Yusef: No, he my brother-in-law.

B Listen and check your answers. Then practise reading the dialogue in groups of three.

C Choose three possible responses for each of the questions below.

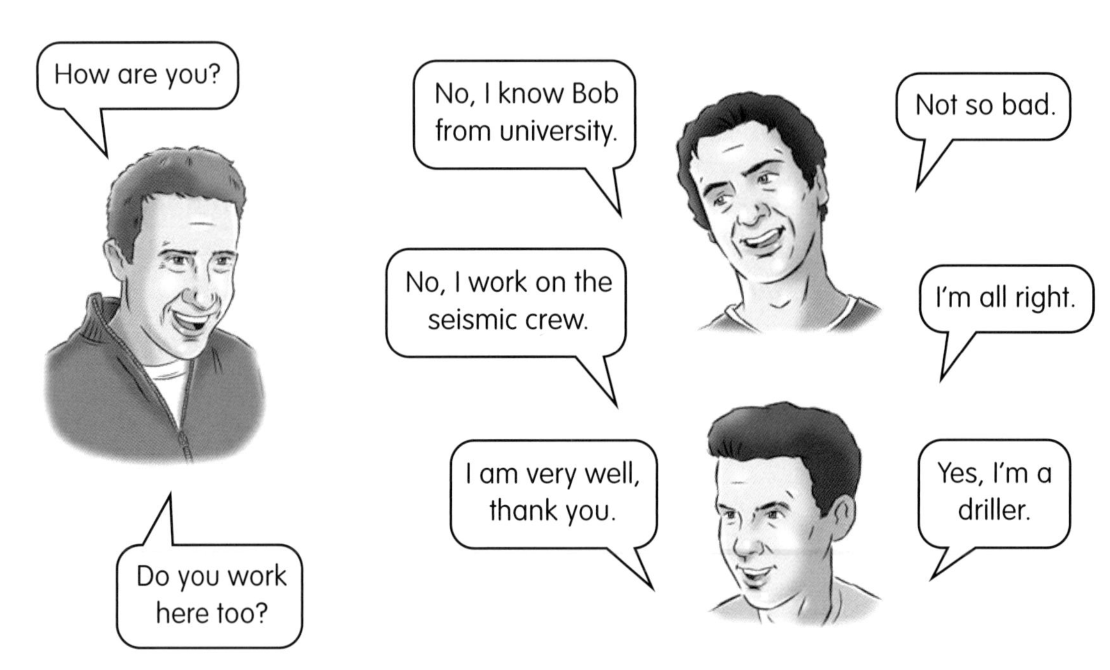

D Match the greetings and questions with the correct responses. Decide if each one uses formal or informal language.

1 How's it going?	How do you do?
2 Let me introduce my colleague, Vasily.	No, I was here last year.
3 How do you do?	Good to meet you.
4 Sorry I'm late. I was held up.	They're all well, thanks.
5 How are your family?	Fine, thanks.
6 Haven't we met before?	I've been on holiday.
7 Is this your first visit to Azerbaijan?	That's okay.
8 What have you been up to recently?	I don't think so.

E Work with a partner. Write your own dialogues using similar questions and responses.

Lesson 3: Asking questions

A Write questions for the responses below.

1 ______________________________? My name's Bob.

2 ______________________________? I'm 32.

3 ______________________________? Yes, I am.

4 ______________________________? It's Helen.

5 ______________________________? Yes, we have two.

6 ______________________________? Paul and Emma.

7 ______________________________? I live in Azerbaijan.

8 ______________________________? I'm a trainee operator.

B Listen and check your answers.

C Choose the correct options to complete the paragraph.

Our / My names are Yusef and Ahmed. I'm 26 and Ahmed *is / are* 34. We *is / are* from Algeria. I *am / is* single. Ahmed *am / is* married to *my / her* sister. They *has / have* three children, and *your / their* second son is called Yusef, too. We *live / lives* in Algiers and we both *work / works* on an oil rig. Ahmed *am / is* a diver, so he *work / works* under water. I'm a derrick monkey at the top of the rig – I *don't / doesn't* like going in the water!

D Complete the tables.

Present simple				
Subject	***be***	***have***	***work***	***like***
I				
you				
he/she/it				
we/they				

Subject	**Possessive adjective**
I	
you	
he	
she	
it	
we	
they	

E Now write a paragraph about yourself.

Lesson 4: Asking for clarification

A **Rearrange the words to make sentences and questions. What do you say if …**

1 someone speaks too quickly?
please / could / speak / you / slowly, / more / ?

Could you speak more slowly, please?

2 you don't understand?
understand / sorry, / don't / I / .

____

3 you didn't hear what someone said?
repeat / could / you / please / that, / ?

____

4 you don't understand an English word?
mean / does / what / 'extinguish' / ?

____

5 you want to know the spelling of a word?
'extinguish' / spell / you / do / how / ?

____

6 you don't know the answer to a question?
know / sorry, / I / don't / .

____

B **Listen and check your answers.**

C **Work with a partner. Match each word or phrase with its meaning.**

1 soiled	to breathe in air, smoke or gas
2 to inhale	to watch or measure an activity
3 to expand	a danger or chance
4 a flame	dirty
5 flexible	to become bigger in size or number
6 a risk	to press something, or make it smaller
7 to monitor	when something can change or move easily
8 to compress	hot, bright, burning gas that we see when something is on fire

D Ask each other the meaning of the words in exercise C.

E Choose the correct option to complete each sentence.

1 It is dangerous to take *risks* / *flames* on the rig.

2 The metal *compressed* / *expanded* when it was heated.

3 We need to *monitor* / *expand* the situation carefully.

4 *Soiled* / *Compressed* air is used to power drills and motors.

5 Plastic is more *soiled* / *flexible* than concrete.

6 He went to hospital because he *expanded* / *inhaled* some dangerous gas fumes.

F Complete the dialogue.

A: Be careful not to inhale the gas fumes.

B: Sorry, I ______________________ understand.

A: The gas – don't inhale it.

B: What? ______________________ repeat ______________________?

A: Don't inhale the gas!

B: ______________________ 'inhale' mean?

A: It means ______________________.

B: How ______________________ spell ______________________?

A: I-N-H-A-L-E.

G Work with a partner. Practise the dialogue.

Lesson 5: Making sentences

Parts of speech

Every word is a part of speech and has a special name:
Nouns are names of things, people, places, ideas and feelings.
Verbs describe actions, processes and states.
Adjectives describe a noun.

Noun	Verb		Adjective	Noun
Divers	*work*	*in*	*deep*	*water.*

A **Complete the table with the words in the box.**

exact equipment operator learn difficult expand flexible monitor job course safe pipeline dangerous calculate extinguish

Noun	Verb	Adjective

B **Look at the other parts of speech. Match each term with its definition.**

1 articles: *a*, *an*, *the*
2 conjunctions: *and*, *but*, *so*
3 adverbs: *well*, *quickly*, *sometimes*
4 prepositions: *in*, *on*, *from*, *to*
5 auxiliary verbs: *have*, *be*, *do*, *will*
6 pronouns: *he*, *you*, *they*, *mine*

- They join one clause to another.
- They are used with a main verb to change the tense, or form a question.
- They give more information about a verb, such as time, place or manner.
- They go in front of nouns to show whether they are definite or indefinite.
- They often replace a noun.
- They link nouns to other elements, such as place or time.

C Identify the parts of speech in the sentences.

1 The equipment is dirty and dangerous.

2 Bob speaks English well.

3 The floorman always works at the bottom of the rig.

4 He is a good, careful worker.

5 The supervisor makes sure that everyone follows the safety regulations.

6 I go to college now, but I will finish in June.

Sentence structure

A standard sentence in English often has the following form:

Subject	**Verb**	**Object**
Bob	*likes*	*his job.*

D Rearrange the words to make sentences.

1 gas / pipelines / transport / .

____

2 the / drillers / drill / operate / .

____

3 derrick / monkeys / under / water / work / don't / .

____

E Look at the forms of the questions. Choose the correct option to complete the rule.

Auxiliary verb	**Subject**	**Verb**	**Object**
Do	*you*	*like*	*Scotland?*
	Verb	**Subject**	
Where	*is*	*Aberdeen?*	

Question forms

In a question, the verb or auxiliary verb comes *before* / *after* the subject.

F Rearrange the words to make questions.

1 rig / work / you / do / on / a / ?

____

2 my / where / overalls / are / ?

____

3 this / mean / word / does / what / ?

____

Lesson 6: Identifying equipment

A Work with a partner. Ask each other questions about the equipment and devices below.

hard hats	chain saw	pig	socket	bolts	gauge
rig	crane	pipeline	plug	forklift	overalls

B **Letters in the English alphabet use certain sounds. Look at the table below. Group the letters in the correct phonetic column.**

~~A~~	B	C	D	E	F	G	~~H~~	I	J	K	L	M
N	O	P	Q	R	S	T	U	V	W	X	Y	Z

/eɪ/ (*play*)	/iː/ (*see*)	/e/ (*men*)	/aɪ/ (*my*)	/əʊ/ (*go*)	/ɑː/ (*car*)	/uː/ (*do*)
A H						

C **Listen and check your answers.**

D **Work with a partner. Practise spelling the equipment and devices on page 12.**

How do you spell 'socket'?

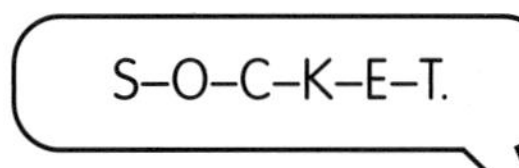

E **The sentences below contain spelling mistakes. Underline the words with mistakes and rewrite them correctly.**

1 Put the plug into the sockit on the wall.

____

2 We bought fourty new hard hats and ten pairs of overals.

____

3 John is a trainee operater and Jim is a mechanic.

____

4 This is a dificult instruement to use.

____

Lesson 7: Talking about opposites

A **Match the words in the box with the pictures.**

smooth	accurate	shallow	approximate		
contract	deep	flexible	expand	rigid	~~rough~~

1 rough

2 ______

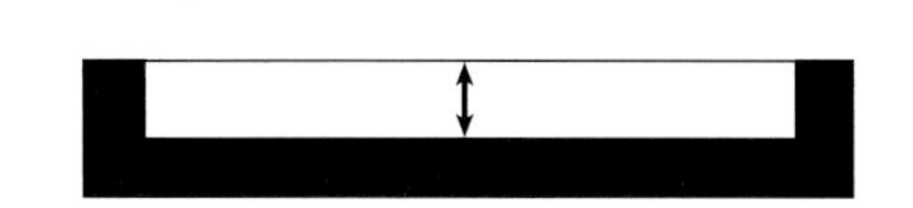

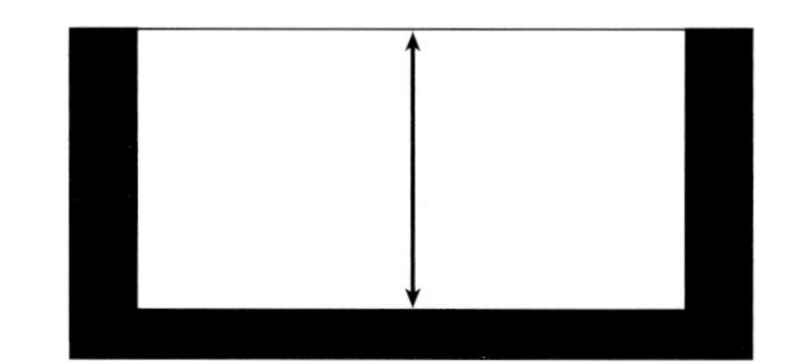

3 ______

4 ______

5 ______

6 ______

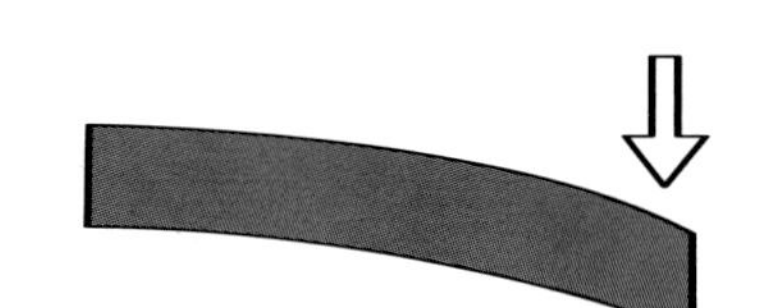

7 ______

8 ______

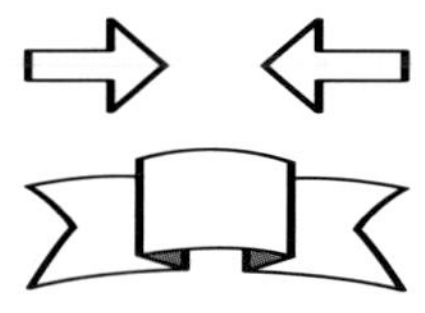

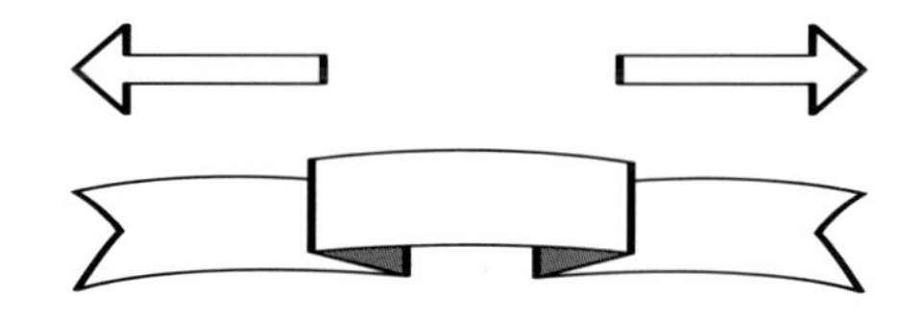

9 ______

10 ______

B **Choose the correct option to complete each sentence.**

1 Iron is a *rigid* / *flexible* material.

2 When you heat metal, it *expands* / *contracts*.

3 It is important to be *approximate* / *accurate* when measuring.

4 A 20-metre well is *deep* / *shallow*.

5 Glass is *smooth* / *rough*.

C **Write more sentences using the words you did not use in exercise B.**

____

____

____

____

____

D **Work with a partner.**
Student 1: Look at the words in box A.
Student 2: Look at the words in box B.

Ask each other questions.

A

heavy	fast	long	early	careful
light	______	______	______	______
big	cool	difficult	flammable	sad
______	______	______	______	______

B

good	dark	loud	dirty	complex
______	______	______	______	______
dry	hot	dangerous	wrong	fragile
______	______	______	______	______

E **Write the opposites of the adjectives in exercise D.**

Lesson 8: Talking about shapes and sizes

A **Complete the table.**

Shape	Noun	Adjective
1 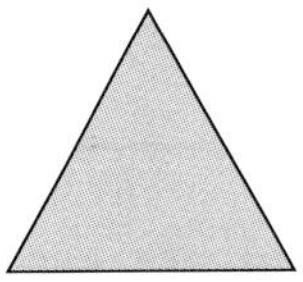	a triangle	______________
2	a rectangle	______________
3	______________	square
4	a circle	______________
5 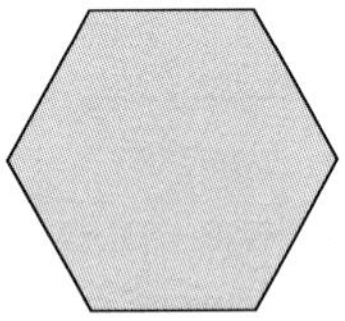	______________	______________
6 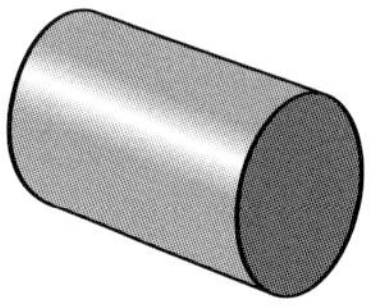	______________	cylindrical
7 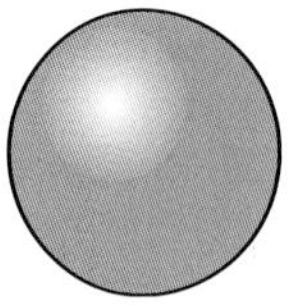	______________	______________
8 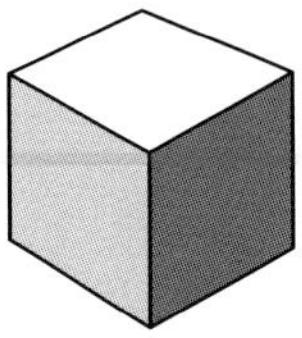	______________	cuboid

B **Listen and check the pronunciation of your answers.**

C **Discuss in groups. What can you see that is ...**
- **round?**
- **rectangular?**
- **cylindrical?**

D **Match each adjective with its opposite.**

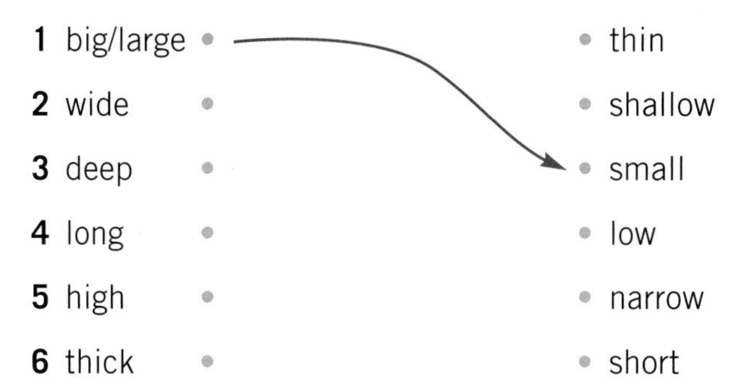

1 big/large	thin
2 wide	shallow
3 deep	small
4 long	low
5 high	narrow
6 thick	short

E **Describe the pictures. Use adjectives from exercise D and the nouns in the box.**

explosion	pipe	hole	pressure	ring

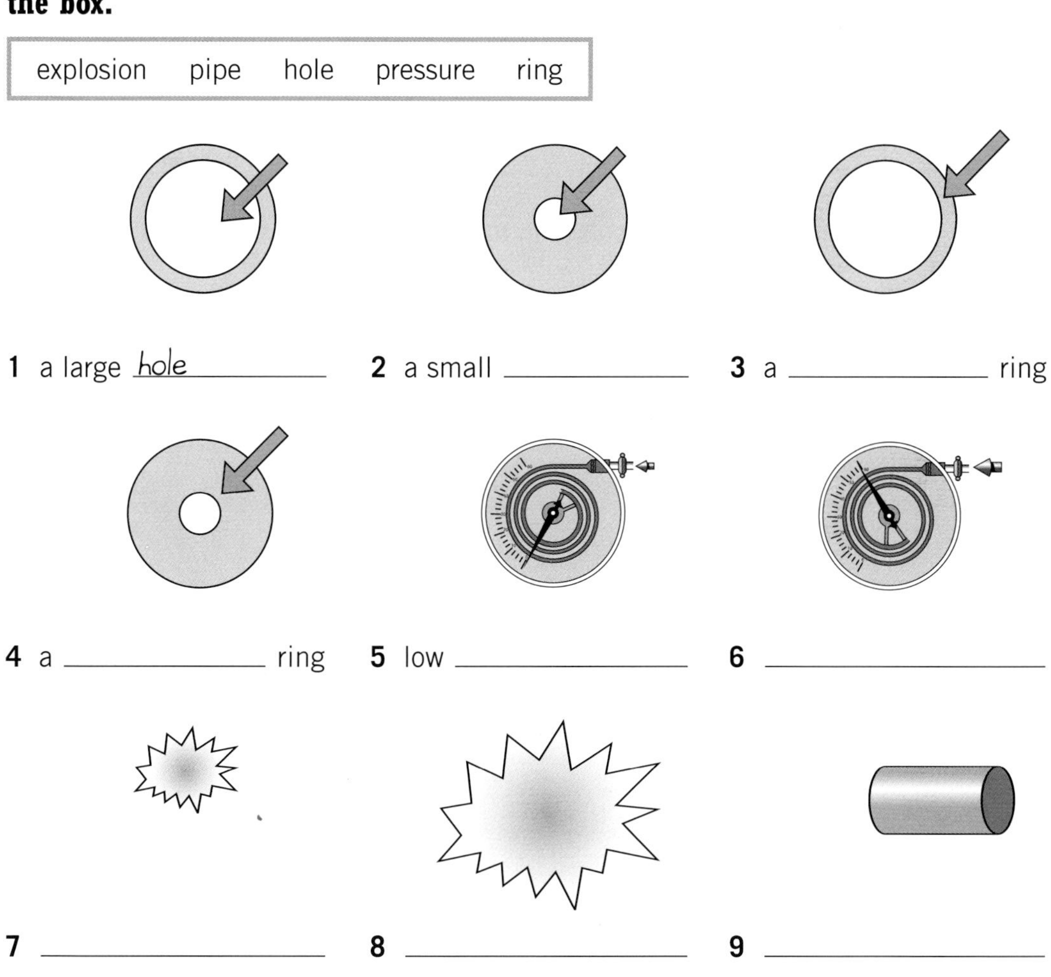

1 a large hole ______
2 a small ______
3 a ______ ring
4 a ______ ring
5 low ______
6 ______
7 ______
8 ______
9 ______
10 ______

Lesson 9: Describing things

A Work with a partner. Try to name six things in your workplace that are ...

1 red ______

2 yellow ______

3 grey ______

4 green ______

5 black and white ______

B Work with a partner. Rearrange the underlined letters to make adjectives.

1 a lgrae edr barge — a large red barge

2 a gnlo nith cable ______

3 a lalt ryeg crane ______

4 dtiry ngeora overalls ______

5 a ckith hetwi pipe ______

Which things can you see in the picture?

C Use the key to complete the sentences and the diagram.

1 The red wire connects terminal A to terminal B.

2 The ______ wire connects terminal B to terminal D.

3 The brown wire connects terminal A to terminal ___.

4 The blue wire connects terminal C to terminal D.

D Work with a partner. Read the definitions and write what they are describing.

1 Black or brown material consisting of clay, water and chemicals: **m** u d

2 A tall, triangular framework over an oil well. It supports equipment, or raises and lowers pipes: **d** __ __ __ __ __ __

3 A large structure where machines are kept for drilling and producing oil and gas. Oil workers live and work here: **r** __ __

4 Cables with detectors that pick up sound waves from under the ground and the ocean: **g** __ __ __ __ __ __ __ __

5 A long, flat or round tank that floats. It can support other structures on water: **p** __ __ __ __ __ __

E Correct the definitions below.

1 A small, square or round, ~~wooden~~ plastic object with two or three metal pins: plug

2 A round, green or pink hat made of plastic: hard hat

3 A thick, rigid, rectangular piece of plastic: credit card

4 A thin, blue, flammable liquid used for fuel: oil

F Work with a partner. Write definitions of three more things. Read them out to another pair. Take it in turns to guess what the definitions refer to.

1 ______________________________

2 ______________________________

3 ______________________________

Lesson 10: Giving definitions

A **Match each job with its definition. Then listen and check your answers.**

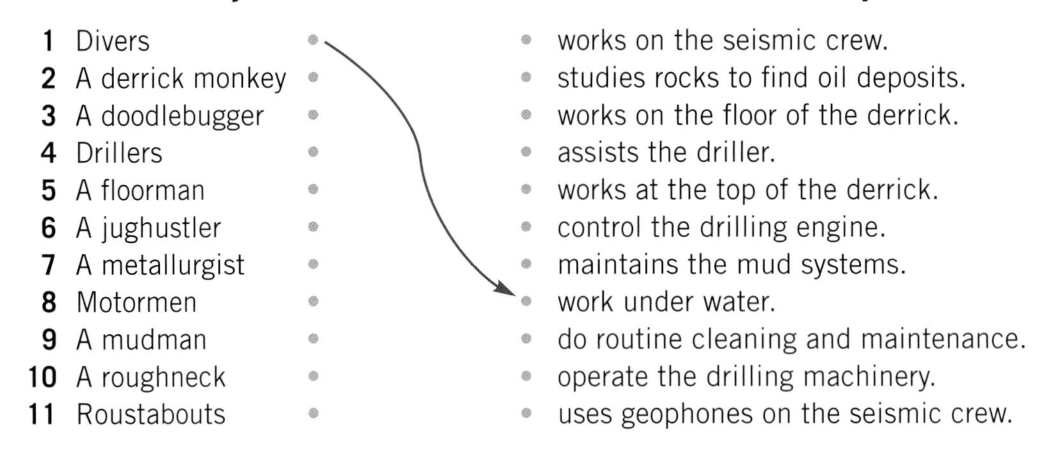

1 Divers
2 A derrick monkey
3 A doodlebugger
4 Drillers
5 A floorman
6 A jughustler
7 A metallurgist
8 Motormen
9 A mudman
10 A roughneck
11 Roustabouts

- works on the seismic crew.
- studies rocks to find oil deposits.
- works on the floor of the derrick.
- assists the driller.
- works at the top of the derrick.
- control the drilling engine.
- maintains the mud systems.
- work under water.
- do routine cleaning and maintenance.
- operate the drilling machinery.
- uses geophones on the seismic crew.

B **Look at the bottle-type submersible platform below. Label it with some of the jobs from exercise B to indicate where the people work.**

C Read the information and identify the types of rig shown in the picture.

There are many kinds of drilling rig suitable for different conditions. What kind of rig is used depends on factors such as how deep the water is.

1 Fixed platform
This kind of rig is suitable for deeper waters (usually 50–300 feet). It is a permanent structure with the drilling rig installed on an underwater jacket (steel structure).

2 Barge-type
This kind of rig is a flat-bottomed barge, suitable for shallow waters. The derrick is over a moon pool in the centre of the barge.

3 Semi-submersible
This kind of rig is suitable for deep-water operations (usually 200–1,500 feet). The rig is supported by floating pontoons that are under the water.

4 Tension leg platform
This kind of rig is similar to a semi-submersible rig, but it is attached to the ocean floor by tensioned steel cables.

5 Jack-up
This kind of rig is suitable for shallow to medium-depth waters. It has supporting legs that can be raised, or 'jacked up', when it moves to another location.

a ______________ b ______________ c ______________ d ______________ e ______________

UNIT 1

Review: Giving basic information

A Find the words in the wordsearch.

~~mud~~
shallow
contract
roustabout
extinguish
socket
seismic
hexagonal
crane
hard hat
explosion
platform
derrick
monitor
fixed

A	R	T	D	S	P	W	C	D	G	H	K
M	T	H	E	O	R	E	T	E	C	A	L
R	Y	P	L	A	T	F	O	R	M	R	W
O	Z	S	E	A	N	E	R	R	U	D	S
U	E	X	T	I	N	G	U	I	S	H	E
S	X	A	I	D	I	B	T	C	H	A	I
T	P	L	Y	F	P	N	W	K	J	T	S
A	L	X	B	H	M	U	D	A	E	N	M
B	O	S	P	F	L	W	B	T	N	S	I
O	S	H	P	I	E	N	T	E	M	O	C
U	I	A	N	X	I	D	C	R	A	N	E
T	O	L	O	E	A	I	N	D	B	S	C
R	N	L	N	D	M	O	N	I	T	O	R
A	B	O	I	B	J	E	C	T	I	C	E
B	D	W	T	B	N	Z	T	S	L	K	R
S	K	C	O	N	T	R	A	C	T	E	R
A	F	I	R	D	A	H	L	L	Y	T	N
N	S	H	E	X	A	G	O	N	A	L	D

B Find words in the wordsearch to match the definitions below.

1 The opposite of 'deep': (adjective) ________________

2 Something you wear to protect your head: (noun) ________________

3 To watch something and make sure there are no problems: (verb) ________________

4 Someone who assists the driller: (noun) ________________

5 It describes the shape of something with six sides: (adjective) ________________

6 The opposite of 'expand': (verb) ________________

C Choose the correct options to complete the paragraph. Then listen and check your answers.

John is a roughneck on an oil rig. He works *on the floor / at the top* of the rig and helps the driller operate the *drilling machinery / geophones.* He wears a hard hat and *overalls / a uniform.* The rig is a semi-submersible rig that is situated in *deep / shallow* waters. It is supported by *a barge / pontoons.*

UNIT 1

Assess your skills: Giving basic information

Complete (✓) the tables to assess your skills.

I can ...	Difficult	Okay	Easy
• ask for and give personal information.			
• ask the trainer for clarification.			
• greet people and respond when meeting them for the first time.			
• name and spell different types of equipment.			
• describe objects and equipment using colour, shape and size adjectives.			
• understand descriptions of equipment, processes and places in the oil industry.			
• explain some oil industry jobs.			

I understand ...		Difficult	Okay	Easy
• the unit grammar:	present simple			
	possessive adjectives			
	terms for parts of speech			
	sentence structure			
• the unit vocabulary (see the glossary)				

If there is anything you are not sure of, ask your trainer to revise the material.

UNIT 2 CALCULATING AND MEASURING

The aims of this unit are to:

- provide the opportunity to practise a full range of number forms used in the technical sphere
- be able to express different types of measurement

By the end of this unit, you will be able to:

- read and use large numbers and make calculations
- talk about common units of measurement
- calculate dimensions for two- and three-dimensional shapes
- understand texts that describe large quantities and measurements

Lesson 1: Saying numbers

1 31/12/06

2 2020

3 1,389

4 33.3%

5 0044 208 250 0403

6 1066

7 0775 176 3298

8 £1,000,000

9 11.35

10 231 m²

A Work with a partner. Take it in turns to say the numbers above. Decide what they refer to.

B Listen and check your pronunciation of the numbers.

C Write the following numbers as words.

1 1 ________

2 12 ________

3 123 a/one ________ and ________-________

4 1,234 one ________, ________ ________ and ________-________

5 12,345 ________________________________

6 123,456 ________________________________

7 1,234,567 ________________________________

8 12,345,678 ________________________________

D Work with a partner.
Student 1: Look at the numbers on page 215.
Student 2: Look at the numbers on page 225.

Take it in turns to say the numbers. Write down the numbers that your partner says. Student 1 starts first.

1 ________________ 4 ________________

2 ________________ 5 ________________

3 ________________ 6 ________________

E Complete the sentences as quickly as possible. Check your answers with a partner.

1 There are 365 ________ days in a year.

2 There are ________ elements in the periodic table.

3 There are ________ cents in a euro.

4 There are ________ seconds in an hour.

5 There are ________ inches in a foot.

6 There are ________ corners in a cube.

7 There are ________ corners in a pyramid.

8 There are ________ elements in pure water.

9 There are ________ consonants in the English alphabet.

10 There are ________ grams in a pound.

Lesson 2: Talking about dates and times

A **You can tell the time in two different ways. Complete the table.**

UK standard time		International time
twelve o'clock	12.00	twelve o'clock
ten past twelve		twelve ten
quarter past twelve	12.15	
	12.25	twelve twenty-five
half past twelve		twelve thirty
twenty to one	12.40	
		twelve forty-five
	12.50	

B **Listen and check your answers.**

C **Work with a partner. Ask each other questions about the activities in the box.**

get up | start work | arrive home | go to bed | have lunch | finish work

What time do you finish work?

At six o'clock.

D **Listen and write the times you hear.**

E **We say different types of numbers in different ways. Choose the correct options to complete the rules.**

Telephone numbers

Say the numbers *individually / in pairs.*

427586 = four, two, seven, five, eight, six
586970 = five, eight, six, nine, seven, zero

Years

Say the numbers *individually / in pairs* (*except for the first ten years of a century).

1985 = nineteen eighty-five
2010 = twenty ten
1901 = nineteen oh one
*2006 = *two thousand and six*

Dates

Say the *day / year*, the name of the month, then the *day / year*.

11/03/99 = the eleventh of March, nineteen ninety-nine
04/10/01 = the fourth of October, two thousand and one

F Write down five important numbers. They should be dates, times and telephone numbers. Ask another student about their numbers.

Why is that date important?

That's when I was born.

____

____

____

____

____

Lesson 3: Talking about fractions and percentages

A **Say the following numbers and write them as words.**

1 ½ a half ______	• 20% ______
2 ¾ ______	• 50% ______
3 ⅓ ______	• 75% ______
4 ⅕ ______	• 87.5% ______
5 ⅞ ______	• 33.3% ______

B **Now match each fraction with a percentage. Check your answers with a partner.**

C **Answer the questions using fractions or percentages.**

1 How much of your time do you spend sleeping?

2 How much of your time do you spend working?

3 What proportion of your country's population lives in towns and cities?

4 How many people are unemployed?

5 What proportion of children learn English?

D **Read the text below and complete the table.**

The majority of injuries, 36%, are caused by slips, trips or falls, and a quarter of accidents are driving related. A fifth of all injuries involve equipment, with 12% of these injuries caused by using damaged equipment, and the rest caused by using the wrong equipment for the job. Just over a tenth of accidents, 11%, are caused by not using the correct personal protective equipment (PPE), and an additional 8% of the injuries are caused by falling objects.

Cause of injury	Percentage
slips, trips or falls	
driving related	
using faulty equipment	
not using appropriate PPE	
using the wrong equipment	
falling objects	

E Listen and finish labelling the pie chart.

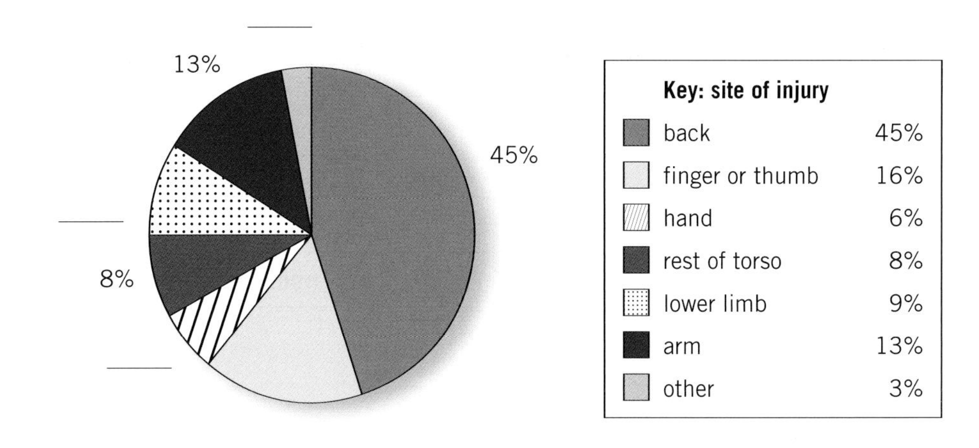

F Talk about the pie chart below. What are the most common types of accidents? Use some of the expressions in the speech bubbles.

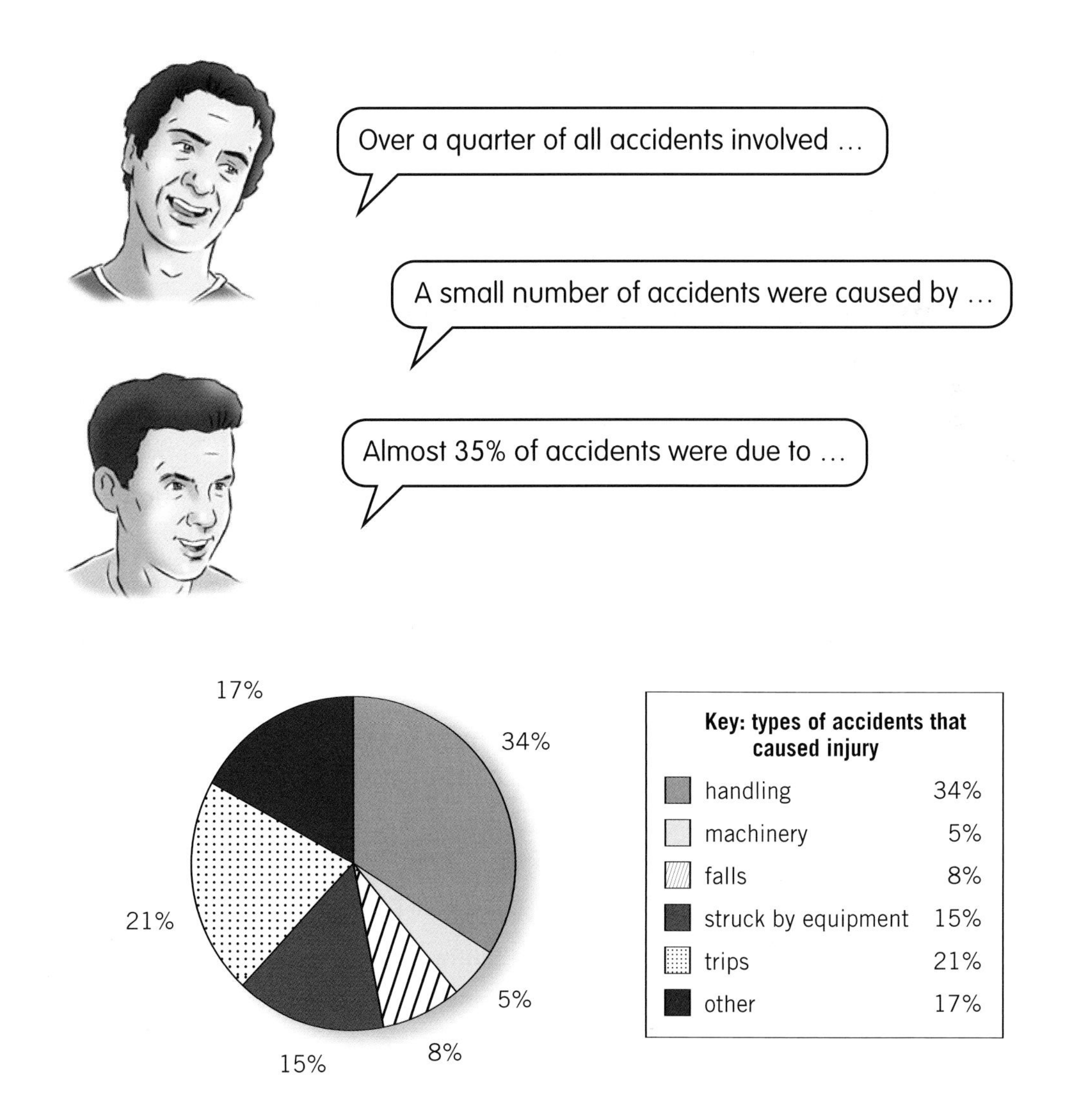

Lesson 4: Talking about nouns

A Complete the table with the words in the box.

~~accident~~ ~~job~~ ~~oil~~ ~~work~~ pipeline gas tank sand filter
engineer water report equipment engine power zinc

Countable nouns	Uncountable nouns
accident/accidents	oil
job/jobs	work

B Add the plural forms to the table.

C Choose the correct option to complete each sentence.

1 I need *an* / *some* information to finish my report.

2 He is very busy – he has *a lot of works* / *a lot of work* to do today.

3 *There are* / *There is* a lot of equipment in the workshop.

4 *How many* / *How much* machines are in the workshop?

5 *There aren't enough* / *There isn't enough* tools in the workshop.

6 *A few* / *A little* accidents are caused by falling objects.

D Some of the words below can only be used with countable or uncountable nouns. Delete the words you cannot use ...

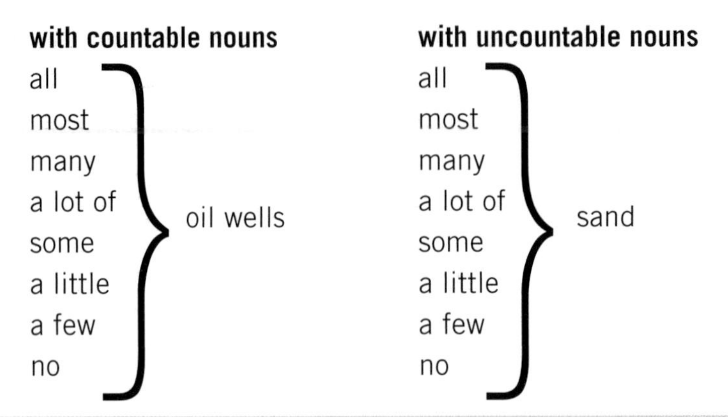

E **Decide whether the words in italics in exercise C are used with countable nouns, uncountable nouns, or both.**

1 ______________________________

2 ______________________________

3 ______________________________

4 ______________________________

5 ______________________________

6 ______________________________

F **Work with a partner.**
Student 1: Look at the seven sentences on page 215.
Student 2: Look at the seven sentences on page 225.

Take it in turns to read your sentences aloud. See if your partner can identify the three sentences which are incorrect.

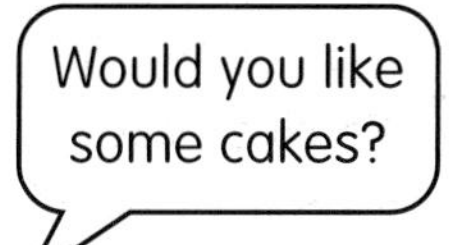

G **Choose the best option to complete each sentence.**

1 *All* / *Some* / *Most* crude oil contains water.

2 *No* / *Most* / *Some* equipment is dangerous.

3 *Most* / *A few* / *Some* chairs have three legs.

4 *A lot of* / *A few* / *Most* men have long hair in my country.

5 Scotland produces *a lot of* / *a little* / *no* oil.

6 *Some* / *A few* / *Very few* people in my organization work 12-hour shifts.

7 *All* / *Some* / *None* of my workmates *has* / *have* had an accident at work.

8 When I write in English, I make *a lot of* / *some* / *very few* mistakes.

Lesson 5: Talking about units of measurement

A **Name the category in each square.**

Weight	Electrical	~~Length~~	Time	Movement	Pressure

1 ____________

km
mm → millimetre
cm ____________
m metre
ft ____________
in inch
____

2 ____________

sec
min ____________
hr minute
____

3 ____________

A
kw ____________
MW hertz
W ____________
Hz watt
V ____________
megawatt

4 ____________

mg milligram
lb ____________
g kilogram
t ____________
kg pound

5 ____________

bar
psi bar
Pa pounds per square inch
kPa pascal
____

6 ____________

kph
mph ____________
rpm miles per hour
fps ____________
revolutions per minute

B **Write the missing words and match each one with its abbreviation.**

C **Practise saying the pairs of numbers below.**

1 **a** 1,500 ft — **b** 500 m

2 **a** 130 mins — **b** 2 hrs

3 **a** 2.9 kg — **b** 4 lbs

4 **a** 78.95 cm — **b** 3 ft 4 ins

5 **a** 240 W — **b** 24 kw

6 **a** 1,000 kph — **b** 700 mph

7 **a** 950 mm — **b** 0.1 m

8 **a** 19 bar — **b** 900 kPa

D **Work with a partner. Decide which number is greater in each pair above.**

E **Listen and check your answers.**

F **Work in a group. Discuss what the following signs mean.**

1

2

3

4

5

6

7

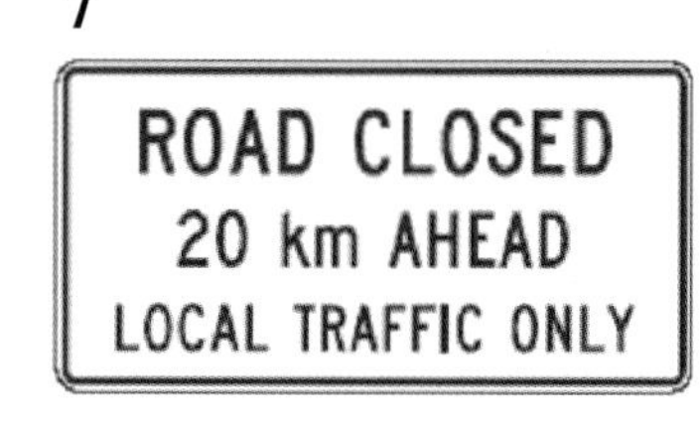

8

Lesson 6: Making calculations

A Match each mathematical symbol with its correct name.

1 + •	• divided by
2 = •	• is not equal to
3 – •	• equals
4 ≈ •	• plus/add
5 x •	• minus/subtract
6 ≠ •	• multiplied
7 ÷ •	• is approximately

B Look at the picture of a calculator. Which letters or symbols are used to mean the following?

1 memory m

2 clear ______

3 squared ______

4 square root ______

5 equals ______

C Write the following calculations in figures. For question 3, you also need to solve the calculation.

1 Fifteen multiplied by seven point three five equals one hundred and ten point two five. ______________________

2 One hundred and twenty-eight divided by twelve point five equals ten point two four. ______________________

3 Twelve divided by four, multiplied by seven, plus eleven equals ______________. ______________________

Area

The area (A) of a rectangle is calculated by the length (L) multiplied by the height H, so L x H = A.

3 cm x 2 cm = 6 cm² (six ***centimetres squared/square centimetres****)*

D Read out the following calculations.

1 11.3 + 18.7 x 6.5 = 195

2 8,500 – 125 ÷ 5 = 1,675

3 1,000 USD = €783

E Look at the rectangles below. Calculate the area of each one.

1 ______________________ 3 ______________________

2 ______________________ 4 ______________________

Which rectangle has the largest area?

Which rectangle has the smallest area?

1 2 cm x 1 cm

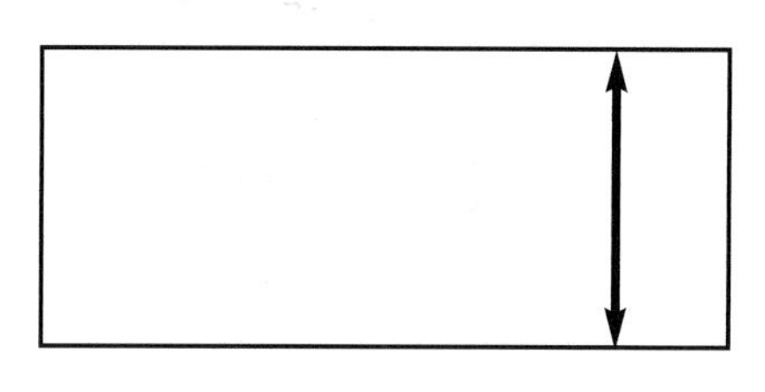

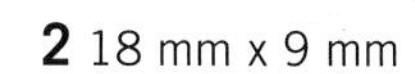

2 18 mm x 9 mm

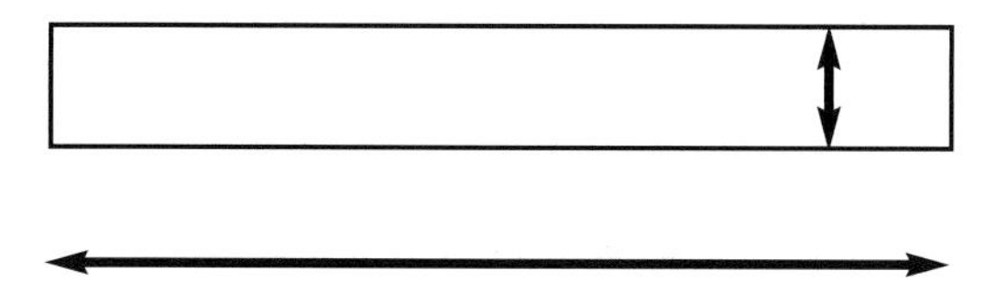

3 22 mm x 3 mm

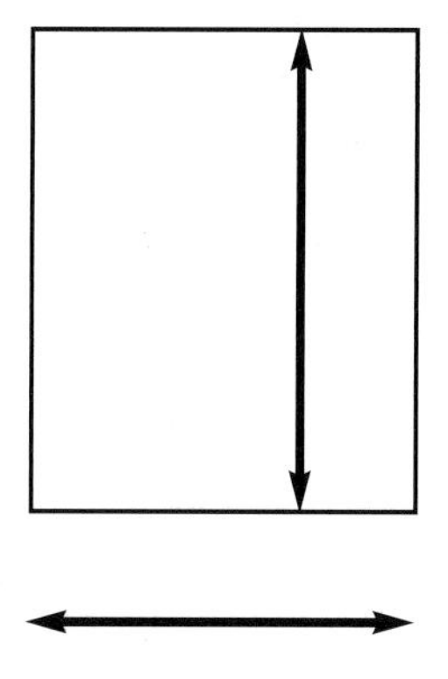

4 10 mm x 12 mm

F Compare your answers with a partner.

What did you make the area of number 1?

I made it two centimetres squared.

Lesson 7: Measuring dimensions

Dimensions

Objects are measured in three dimensions: height, width and length.

A **Label the diagram below with the correct dimensions.**

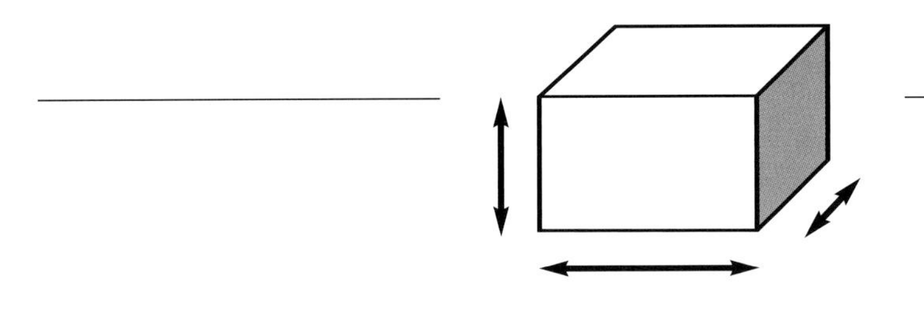

B **We use different adjectives to describe dimensions. Complete the table with the adjectives in the box. Some adjectives can be used more than once.**

wide short tall high narrow shallow deep long

Height	Width	Length

We can describe dimensions in two ways:

- with an adjective

  *The box is ten centimetres **high**.*

- with a noun

  ***The height** of the box is ten centimetres.*

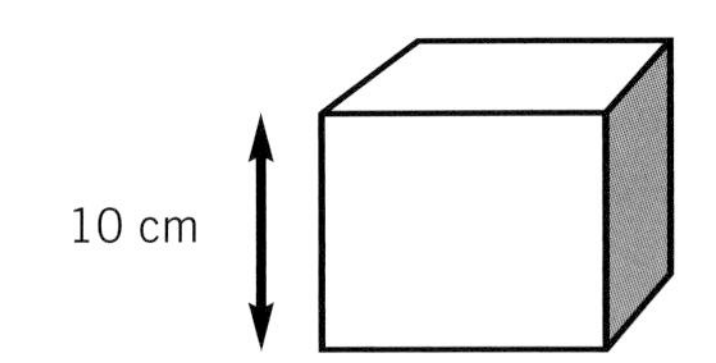

C **Read the description of the diagram below. Choose the correct option to complete each sentence.**

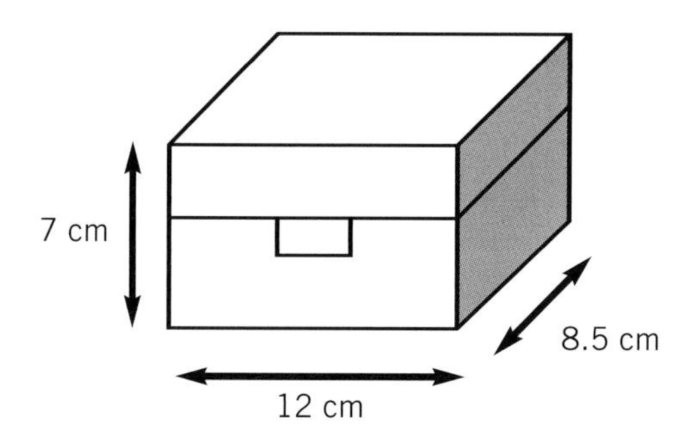

The box is 7 centimetres *high* / *height*.
Its *wide* / *width* is 12 centimetres.
Its *long* / *length* is 8.5 centimetres.

D **Write similar descriptions for the diagrams below.**

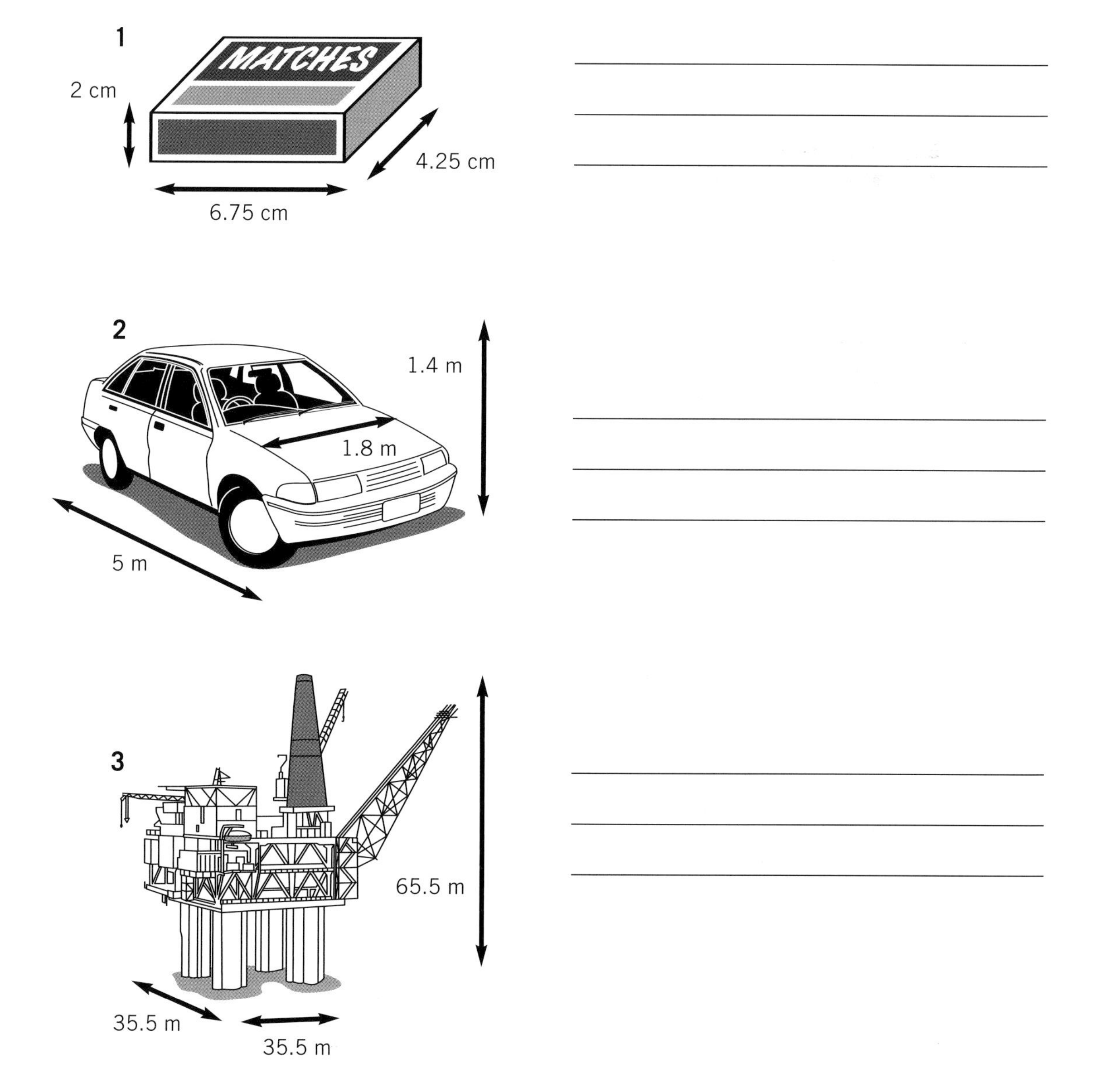

1 ____

____

____

2 ____

____

____

3 ____

____

____

E **Listen and compare the descriptions with yours.**

Lesson 8: Measuring circles and pipelines

A **Read the descriptions and write the names of the parts of a circle. Draw the parts on the diagrams.**

diameter (D)   radius (R)   circumference (C)

Description	Part of circle	Diagram
The length of the outside of a circle.	circumference	1
The distance from one side of a circle to the other, through the centre.		2
The distance from the edge of a circle to the centre.		3

Circumference/Area

The circumference of a circle (C) = $2\pi r$.
The area of a circle (A) = πr^2.

B **Use a ruler to calculate the circumference and the area of the circles above. Check your answers with a partner.**

	Circle 1	Circle 2	Circle 3
circumference			
area			

What did you make the circumference of circle 1?

Volume

To calculate the volume of a pipeline, the area of the internal circle is multiplied by the length of the pipeline = $\pi r^2 \times L$.

The internal diameter of the pipeline below is 1 metre. The length is 4 metres. Therefore, the volume is 3.141 x 0.5 x 4 = 6.282m^3 (six point two eight two **metres cubed/cubic metres**).

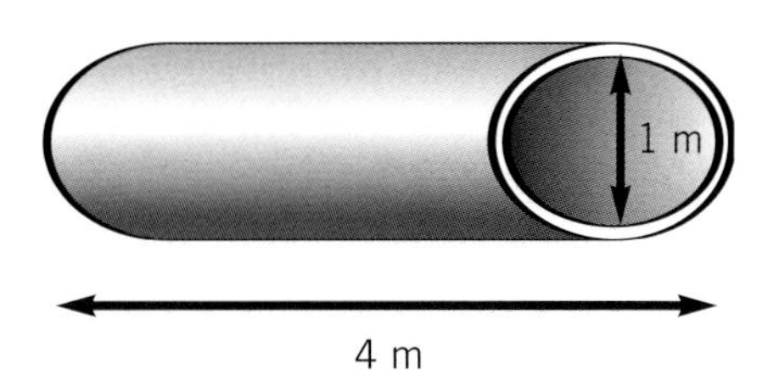

C Use the formula to calculate the volume of the pipelines and barrel below.

1

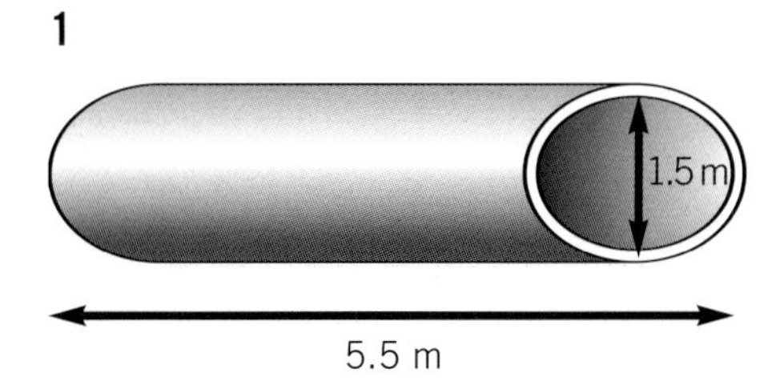

Volume = ______________________

2

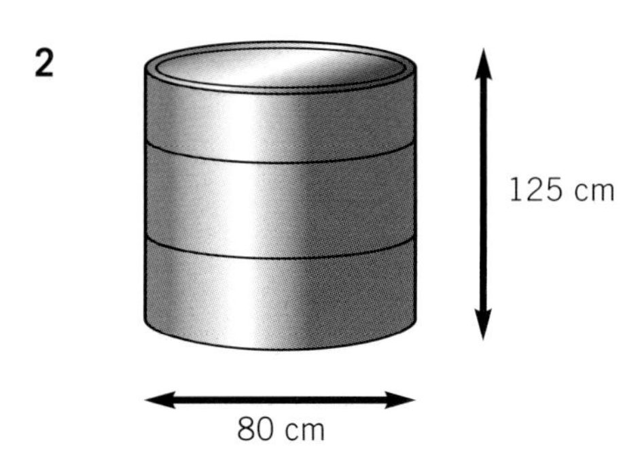

Volume = ______________________

3

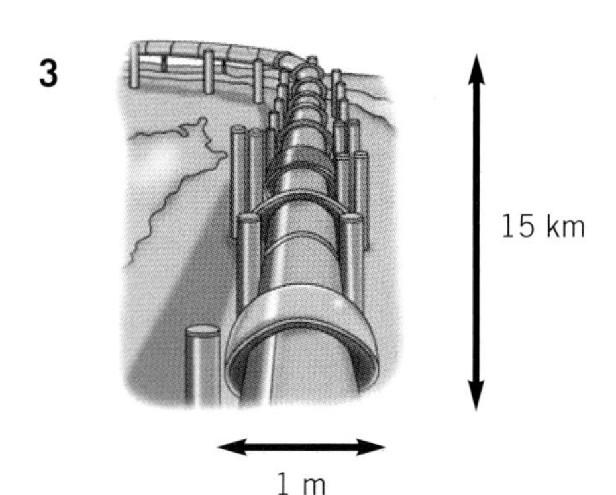

Volume = ______________________

Lesson 9: Taking other measurements

Temperature

Temperature is measured in:
1 degrees Celsius (°C) **2** degrees Fahrenheit (°F) **3** degrees Kelvin (°K)

A Complete the sentences below. Check your answers with a partner.
Student 1: Look at page 215.
Student 2: Look at page 225.

1 The freezing point of water is ____________________.

2 The boiling point of water is ____________________.

3 The normal temperature of the human body is ____________________.

4 The surface of the Sun is ____________________.

5 0°K is ____________________°C.

6 32°F is ____________________°C.

Sound

Sound is measured in decibels (dB).

B Guess how loud these things are. Match them with the decibel levels.

1 noisy factory •	• 200 dB
2 chain saw •	• 120 dB
3 speech at 1 m •	• 100 dB
4 moon rocket at 300 m •	• 80 dB
5 car horn at 4 m •	• 60 dB
6 quiet office •	• 40 dB

C Listen and check your answers.

Distance

Long distances are measured in miles (m) or kilometres (km).
One mile = 1.61 kilometres.

D **Work with a partner. Find the pairs of cities in the table on the map and guess the distance between them.**

Cities	Distance between cities
London–Moscow	
Prague–Mumbai	
Istanbul–Tehran	
Bangkok–Tokyo	
Kuwait–New Delhi	
Cairo–Singapore	
Rome–Dubai	
Paris–Hong Kong	

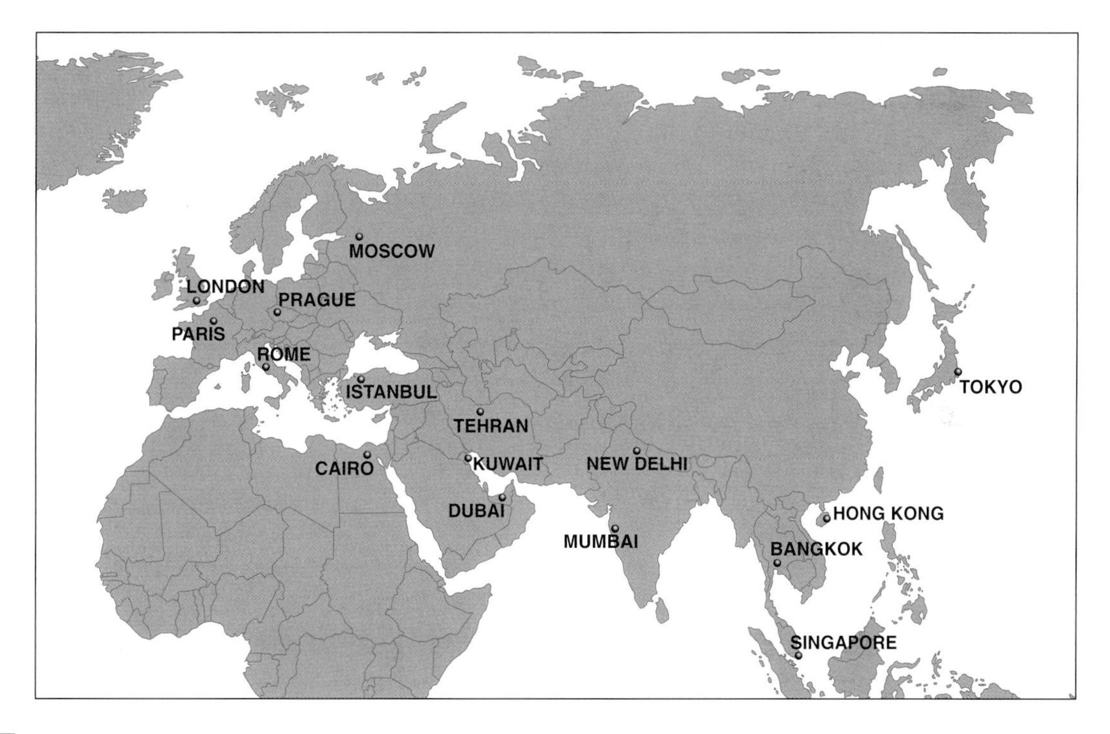

E **Work with a partner.**
Student 1: Look at the answers on page 216.
Student 2: Look at the answers on page 226.

Swap information so that you can complete the right-hand column of the table.

Lesson 10: Measuring pressure

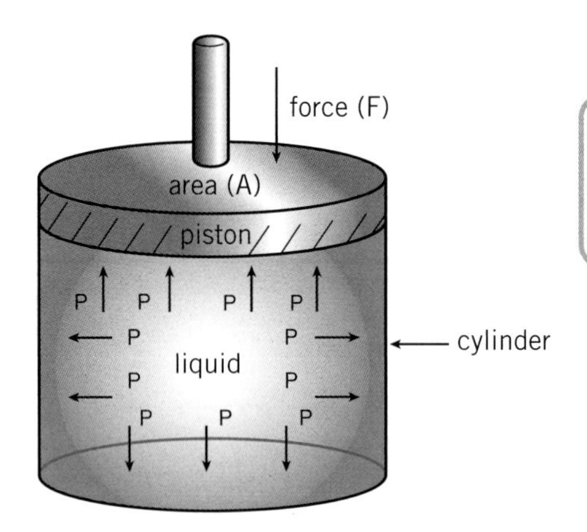

Pressure

Pressure (P) is the force (F) divided by the area (A), so $P = F \div A$.

A **Pressure is measured in different units. Read the text below and complete the table.**

Some companies use Imperial units (USA system) and some companies use International Standard Metric Units (IS). Imperial units measure the force in pounds, and the area in square inches. This gives a pressure in pounds per square inch (psi). International Standard Metric Units measure area in square metres and force in newtons. This gives a pressure measurement called pascal (Pa), which is newtons per square metre. Because this is a very small unit, the kilopascal (kPa) is often used.

	Imperial	ISO
force		
area		
pressure		or

B **Read the formulae below and use them to convert the pressures. Check your answers with a partner.**

Conversion formulae

To convert (change) psi to Pa, the following formula is used: 1 psi = 6895 Pa.

High pressures are measured in bars: 100 kPa = 1 bar.

1 10 kPa = ______________ Pa

2 20 psi = ______________ Pa

3 15 bar = ______________ psi

4 848085 Pa = ______________ psi

5 250 psi = ______________ kPa

6 1 bar = ______________ psi

C Match each word with its definition.

1 reservoir	How easy it is for a liquid to flow.
2 mud	The weight of a unit of a substance.
3 viscosity	A mixture of earth and water.
4 density	A place that stores a large amount of liquid, e.g., an oilfield.

D Read the text and answer the questions.

Drilling mud is a mixture put in a well to control the pressure of the reservoir fluids. It also cleans the drill bit and the bottom of the well, and carries cuttings out of the well. A mud engineer is responsible for the drilling mud. He changes the mixture of the mud to suit the well.

Water-based drilling mud is very common. It contains fresh water or sea water, bentonite clay to give viscosity, and barite to give density. It also contains chemicals that the mud engineer changes for different conditions. To make 800 bbl (barrels) of a water-based drilling mud with a density of 12 lb per US gallon, mix:

fresh water	672 bbl
barite	1552 sacks x 100 lb
bentonite clay	104 sacks x 100 lb
filtration agent	320 lb
co-polymer	120 lb

1 Drill mud does three things. What are they?

____

____

____

2 What does the mud engineer do?

____

3 Does a mud engineer use barite to help the mud flow?

____

4 How many sacks of bentonite are in 400 barrels of water-based drilling mud?

____

5 A mud engineer wants a density of 9 lb per US gallon. What does he do?

____

UNIT 2

Review: Calculating and measuring

A Work in teams. Guess the answers to the questions.

1 **The first element in the periodic table is ...**
a oxygen **b** hydrogen **c** carbon

2 **The population of Britain is ...**
a approximately 60,400,000 **b** approximately 65,000,000
c approximately 55,500,000

3 **When did World War II begin in Europe?**
a 3rd October, 1937 **b** 1st December, 1938 **c** 1st September, 1939

4 **What proportion of native English speakers live in the USA?**
a just over ⅕ **b** just under ¼ **c** just over ⅓

5 **The record for deep-water oil production is ... under the sea.**
a 5,000 ft **b** 1,800 m **c** 2 km

6 **In the USA, 535,000 barrels of gasoline are produced every day. If there are 42 gallons per barrel, how many gallons are produced per day?**

7 **Which countries produce ...**
a approximately 11.77% of the world's oil?
b approximately 10.36% of the world's oil?

8 **Put the distances below in order of size (from small to big).**

centimetre	mile	kilometre	inch	foot	millimetre	metre

9 **Put the times in order of length (from short to long).**

minute	week	season	second	fortnight	hour	year	month	day

10 **Put the quantities in order of size (from small to big).**

250 kPa	6895 Pa	1 bar	20 psi

B Listen to someone describing one of the world's largest oil pipelines. Complete the table with the statistics you hear.

Name of pipeline	Baku-Tbilisi-Ceyhan pipeline (BTC)
Length	
Diameter	
Likely capacity in 2009	

UNIT 2

Assess your skills: Calculating and measuring

Complete (✓) the tables to assess your skills.

I can ...	Difficult	Okay	Easy
• understand and say large numbers.			
• understand and talk about dates and times.			
• understand and say percentages and fractions.			
• understand and talk about countable and uncountable quantities.			
• recognize and convert units of measurement.			
• read and work out basic calculations.			
• measure and describe the dimensions of objects.			
• understand a text that describes quantities.			

I understand ...	Difficult	Okay	Easy
• the unit grammar: countable/uncountable nouns			
• the unit vocabulary (see the glossary)			

If there is anything you are not sure of, ask your trainer to revise the material.

UNIT 3

DESCRIBING EQUIPMENT

The aim of this unit is to:

- provide the language and skills necessary to describe a range of common oil industry hand tools

By the end of this unit, you will be able to:

- identify a range of different hand tools
- express ability using *can*
- describe the location of objects with accuracy
- identify the parts of common hand tools
- describe the relationships between the parts of common hand tools

Lesson 1: Talking about workshop tools

A Work with a partner. Discuss which tools you have seen or used.

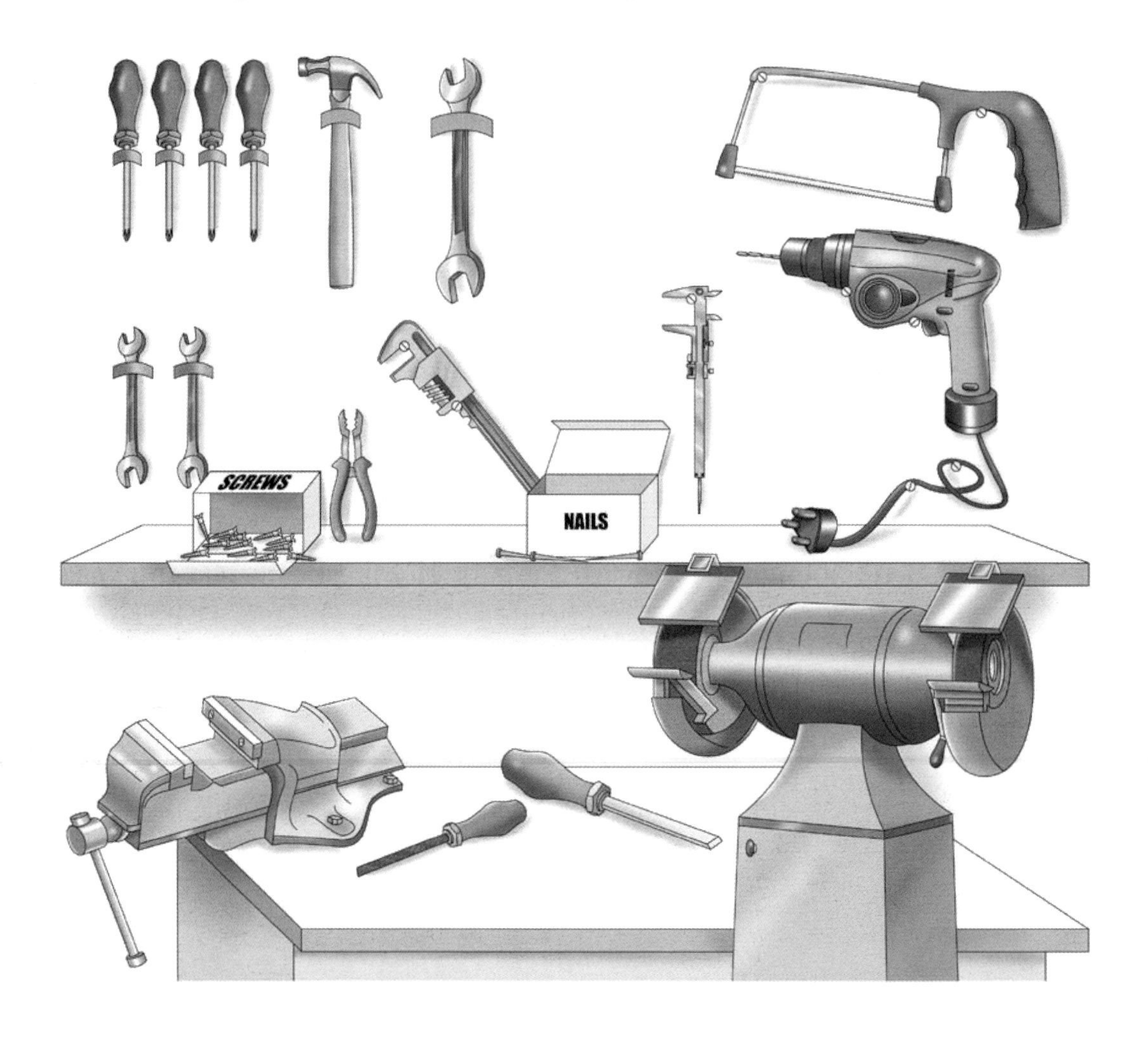

B Look at the diagrams and match each one with a verb in the box.

tighten	chip away	grip	sharpen	loosen

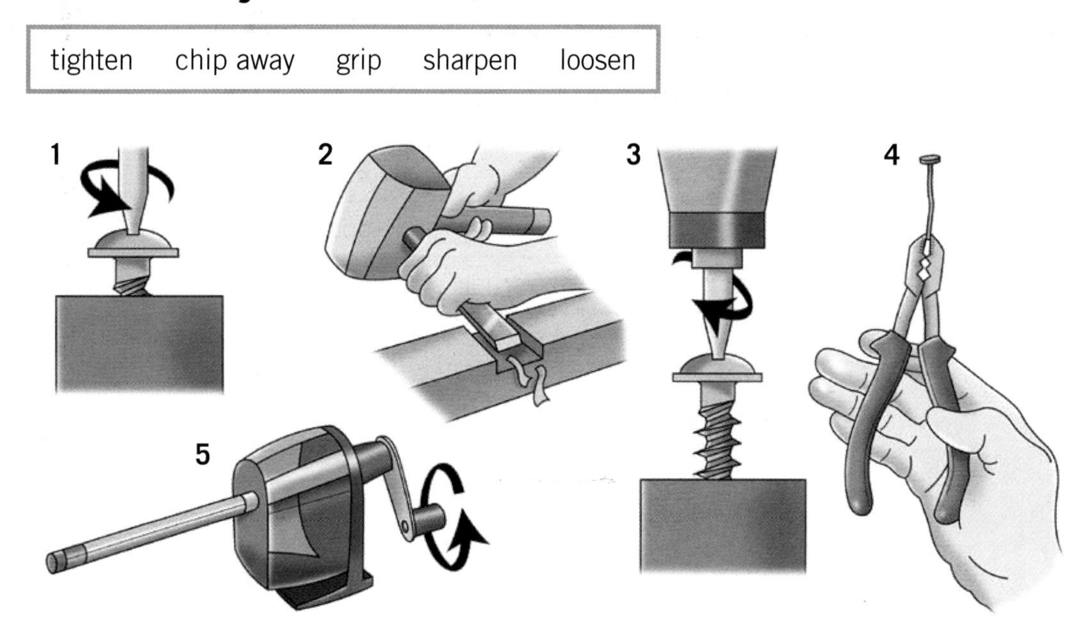

C Match each tool with its use.

You can use …

1 a grinder	to cut a piece of wood or metal.
2 a screwdriver	to make a hole in a piece of wood or metal.
3 calipers	to hold a piece of wood or metal securely in place.
4 a file	to tighten a screw.
5 a hammer	to rotate a pipe.
6 a saw	to finish the surface of a piece of metal.
7 a spanner	to grip small objects.
8 a pipe wrench	to connect two pieces of wood with a nail.
9 a drill	to sharpen other tools.
10 a chisel	to measure internal or external dimensions.
11 a vice	to loosen a bolt.
12 pliers	to chip away metal.

D Read and complete the description of what you need to make a bench.

To make a bench, you need a saw to [1] ________ the wood, a vice to [2] ________ the wood, a drill to [3] ________ holes in the wood and a screwdriver to [4] ________ the pieces of wood together.

E Listen and check your answers.

F Discuss what you need to …

1 change an electric plug.

2 put a shelf on a wall.

Lesson 2: Expressing ability

A **We use *can/can't* to express ability. Look at the table and complete the sentences.**

	Bob can	**Bob can't**
Ahmed can	use a drill	speak Arabic
Ahmed can't	drive a truck	operate a forklift

1 Bob can ______________________, but Ahmed can't.

2 Bob and Ahmed can both ______________________.

3 Bob and Ahmed can't ______________________.

4 Bob can't ______________________, but Ahmed can.

B **Read the text and underline the five grammatical mistakes.**

My brother's an English teacher and he can read and write English really good. I can to speak English, but I don't can write it. I'm more practical though. I can use a drill and repairing things in the house. My brother no can do that.

C **Now write out the correct version.**

My brother's an English teacher and he ______________________

____

____

I can ______________________

____

____

I'm more practical though. ______________________

____

____

D **Listen and check your answers.**

E **Ask questions and complete a table about yourself and a partner. Use some of the ideas in the box.**

change a fuse	repair electrical problems	operate a crane
fly a plane	drive a van	use a computer	understand heating systems
speak Russian	cook a meal	use electrical tools
do mathematical calculations in my head		

	I can	I can't
________________ can		
________________ can't		

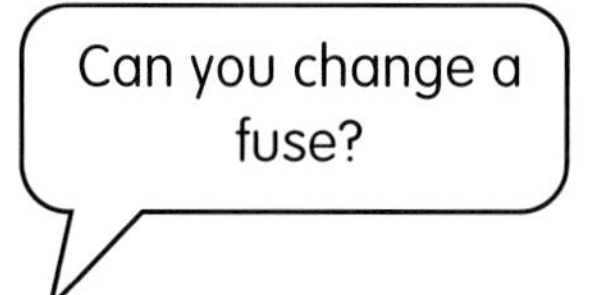

F **Report your findings to the class.**

G **Work in groups. Write sentences about your group using *can* or *can't*.**

A few people in the group can ____.

No one ____.

Some people ________________________________, but others can't.

We can all ____.

Only one person ____.

H **We also use *can/can't* for possibility. Complete the sentences with *can* or *can't*.**

1 You ______________ remove a nail with a file.

2 You ______________ cut electric cable with a pair of scissors.

3 You ______________ cut a pipe with a hammer.

4 You ______________ turn a bolt with pliers.

5 You ______________ chip away metal with a screwdriver.

6 You ______________ finish metal with a saw.

7 You ______________ hold a piece of wood with a vice.

8 You ______________ sharpen tools with a file.

Lesson 3: Describing place and position

A Look at the picture on page 46 and read the sentences below. Choose the correct preposition of place to describe the position of each tool.

1 The hammer is *next to / under* the spanner.

2 The saw is *over / above* the drill.

3 The screws are *on / inside* the box.

4 The pipe wrench is *above / below* the hammer.

5 The spanners are *under / below* the screwdrivers.

6 The nails are *in / in front of* the box.

7 The pliers are *behind / in front of* the screws.

8 The file is *inside / between* the vice and the chisel.

B Make four more sentences about the other tools.

C Finish labelling the diagrams below with words from the box.

bottom	right-hand side	front	middle	corner

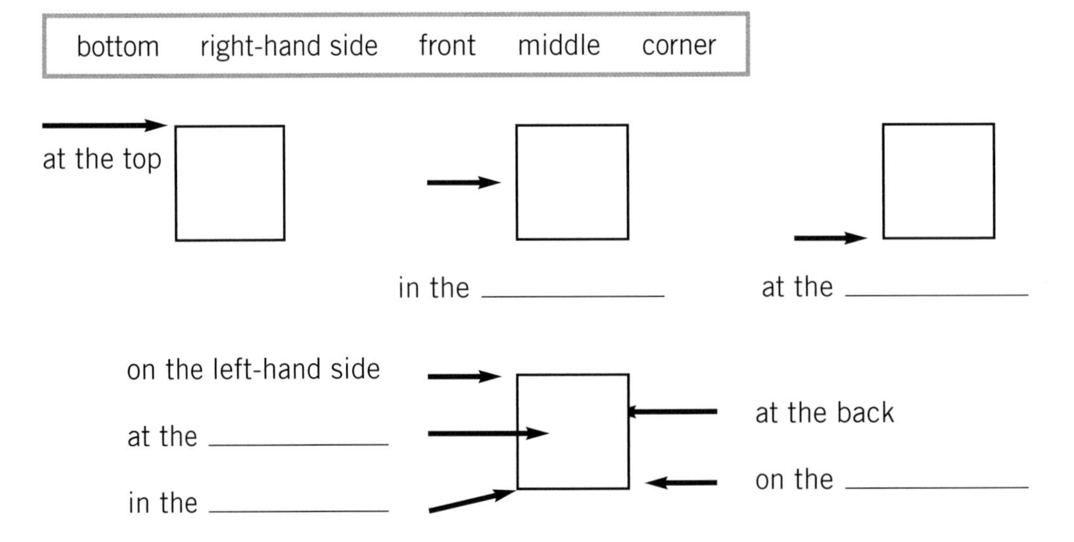

D Work with a partner.
Student 1: Look at the picture of a workshop on page 216.
Student 2: Look at the picture of a workshop on page 226.

Describe the position of the tools. Find eight differences between the pictures without looking at your partner's picture.

Where are the pliers in your workshop?

On the left-hand side, next to the screws.

E **Read the description of a workshop and draw a floor plan of it in the box below.**

There are some steps up to the door, which is on the left-hand side of the workshop. There is a workbench along the back wall, and the air filter and conditioning unit are to the right of this. Opposite the door, near the right-hand wall, is a table saw. There is a planer next to it, with a router table in front.

F **Add four more workshop items to your diagram and explain to a partner where they are.**

Lesson 4: Describing tools

A **Look at the diagram below. Complete the text with the verbs from the box in the correct form.**

use	be	supply	have	connect	hold	control

This [1] __________ an electric drill. You can [2] __________ it to make holes in wood, metal or concrete. It [3] __________ a motor inside it, a trigger, a power cord and plug at the bottom, and a chuck and a bit at the front. The power cord [4] __________ power to the motor. The chuck [5] __________ the bit in place. The trigger [6] __________ the motor. The plug [7] __________ the power cord to the power supply.

B **Listen and check your answers.**

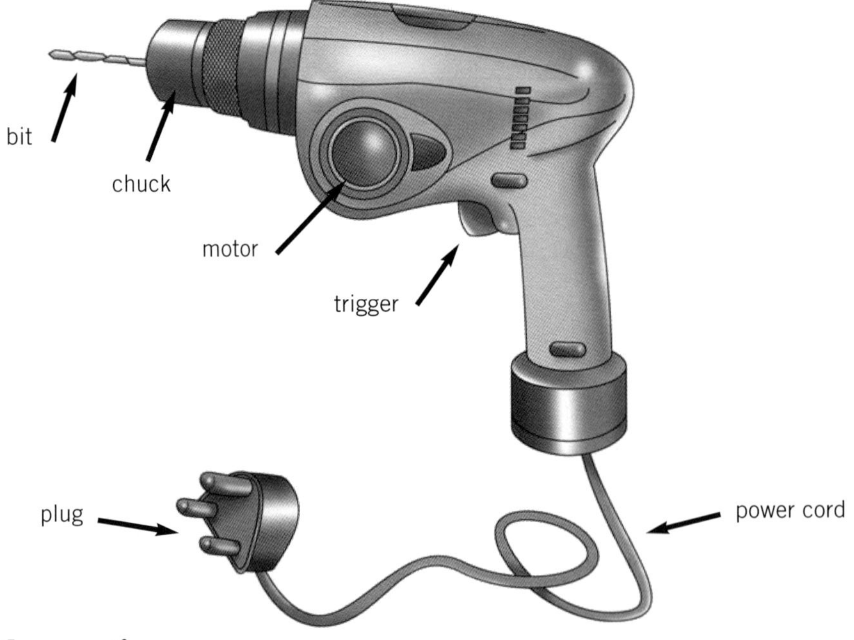

C **Answer the questions.**

1 Where is the bit?

____

2 What does the trigger do?

____

3 Where is the plug, and what does it do?

____

D **Look at the two diagrams below and correct the mistakes in the texts.**

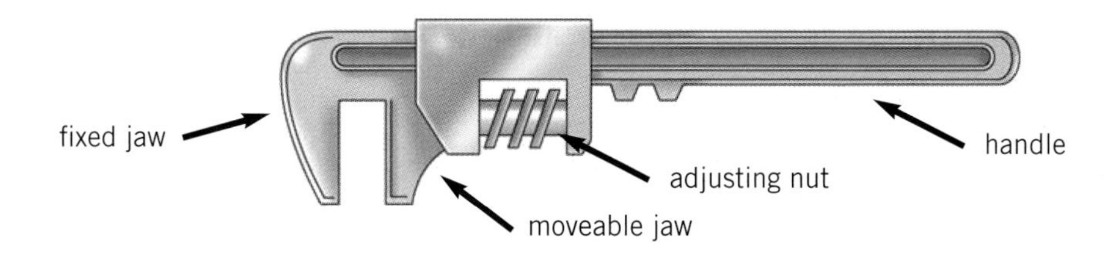

This is a pipe wrench. You can use it to rotate pipes. It has a handle, two adjusting nuts, a fixed jaw and a moveable jaw. The adjusting nut is in front of the moveable jaw and adjusts the position of the fixed jaw.

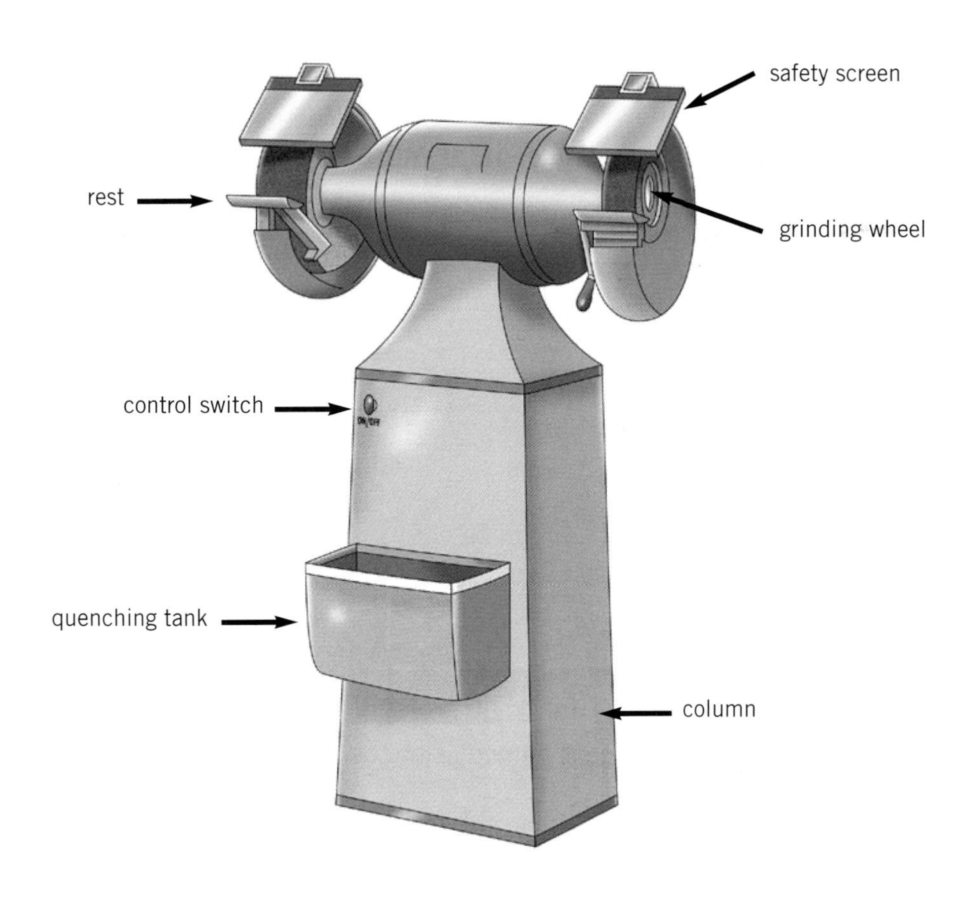

This is an off-hand grinder. You can use it to recondition tools like screwdrivers and chisels. It has a column, a grinding wheel, safety screens, rests, a control switch and a quenching tank. The control switch on the left of the column operates the grinding wheels. The rests above the wheels hold the work in place, and the screens protect the user from debris. The quenching tank behind the column is used to cool the work.

E **Listen and check your answers.**

Lesson 5: Talking about objects

A **Identify the subject, the verb and the object in the following active sentences.**

1 A motor powers the drill.

2 The chuck holds the bit in place.

3 Drillers operate the machines.

4 Ahmed can use an electric drill.

The passive voice

Sometimes in sentences, the subject is not important, or we do not know the subject. In these sentences, we often use the passive voice. The object is moved to the beginning of the sentence and becomes the passive subject.

B **Now put the sentences in exercise A into the passive.**

	Passive subject	Auxiliary	Verb (past participle)	(*by* + agent)
1	The drill	is	powered	by a motor.
2	____________	____________	held (in place)	by the chuck.
3	The machines	are	operated	____________.
4	An electric drill	can be	____________	____________.

C **Complete the rule.**

The passive voice

The passive voice is formed using the auxiliary verb *be* and the ____________ participle.

It is not always necessary to include *by* + ____________ in a passive sentence.

About 25% of sentences in academic texts use the ____________ voice.

D **Identify the passive sentences in the text.**

Grinders are used to sharpen, cut or polish many different materials. They can be used to sharpen other cutting tools in the workshop. Some grinders are attached to a workbench, while others are held in the hand. Safety goggles should be worn by the person who operates the grinder.

E Write the past participle of the words below.

1 use ______________________

2 monitor ______________________

3 measure ______________________

4 make ______________________

5 show ______________________

6 hold ______________________

7 wear ______________________

8 attach ______________________

9 turn ______________________

10 take ______________________

11 connect ______________________

12 drill ______________________

F Rewrite the active sentences below using the passive. Then listen and check your answers.

1 The plug connects the drill to the energy supply.

The drill ______________________ to the energy supply by the plug.

2 The engineer uses a file to finish the metal.

A file ______________________ to finish the metal.

3 We make the holes with an automatic drill.

The holes ______________________.

4 A computer monitors the flow.

The flow ______________________.

5 A pointer shows the change in level.

The change ______________________.

6 You can hold pipes in place with a pipe wrench.

Pipes can ______________________.

7 You can measure liquid in a tank with a float.

Liquid in a tank can ______________________.

8 You cannnot turn the screw clockwise.

The ______________________.

Lesson 6: Describing measuring devices (1): pressure and temperature

A Match each word with its definition.

1 bulb •	• A curve that turns around a central point in the shape of waves.
2 pointer •	• The rounded part of an instrument or vessel.
3 tube •	• A series of numbers or marks at set intervals.
4 spiral •	• A cylinder made of glass, metal, cardboard or plastic.
5 scale •	• A thin piece of metal that points at a scale.

B Read the descriptions of four types of pressure-measuring equipment, then match each description with a picture.

1 A **bellows** is made of thin metal in the form of a cylinder that has deeply corrugated walls. It is connected to a pointer.

2 A **bourdon tube** consists of a thin, metal-walled tube. The bourdon tube is connected to a pointer.

3 A **differential pressure bellows** consists of two bellows, working in opposition, connected to a pointer.

4 A **helical bourdon tube** consists of a bourdon tube in the shape of a spiral, connected to a pointer.

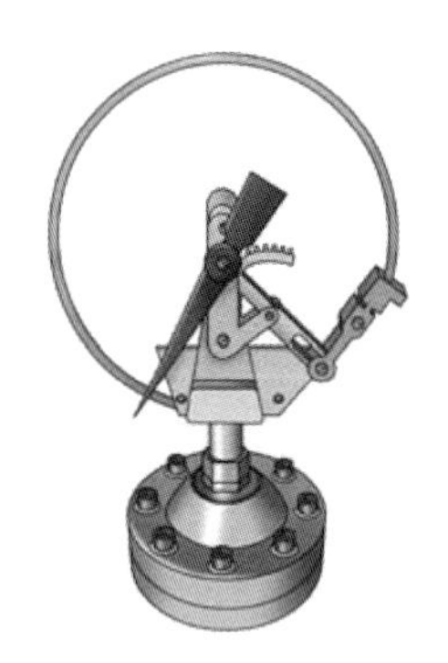

a ______________________

b ______________________

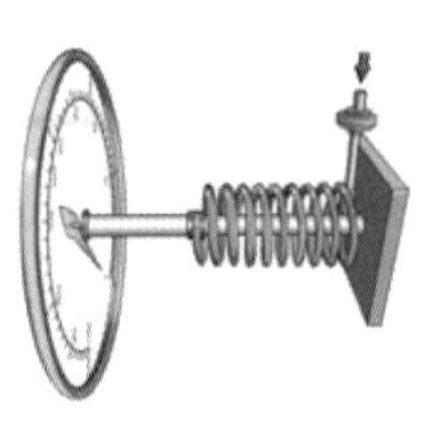

c ______________________

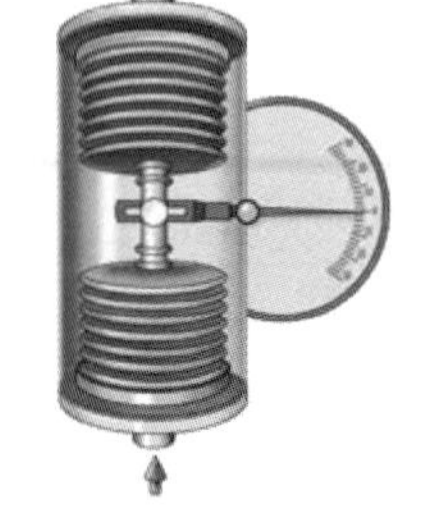

d ______________________

C Label the diagram of a filled system thermometer with the words in the box.

bourdon tube	temperature scale	capillary tube	pointer	bulb

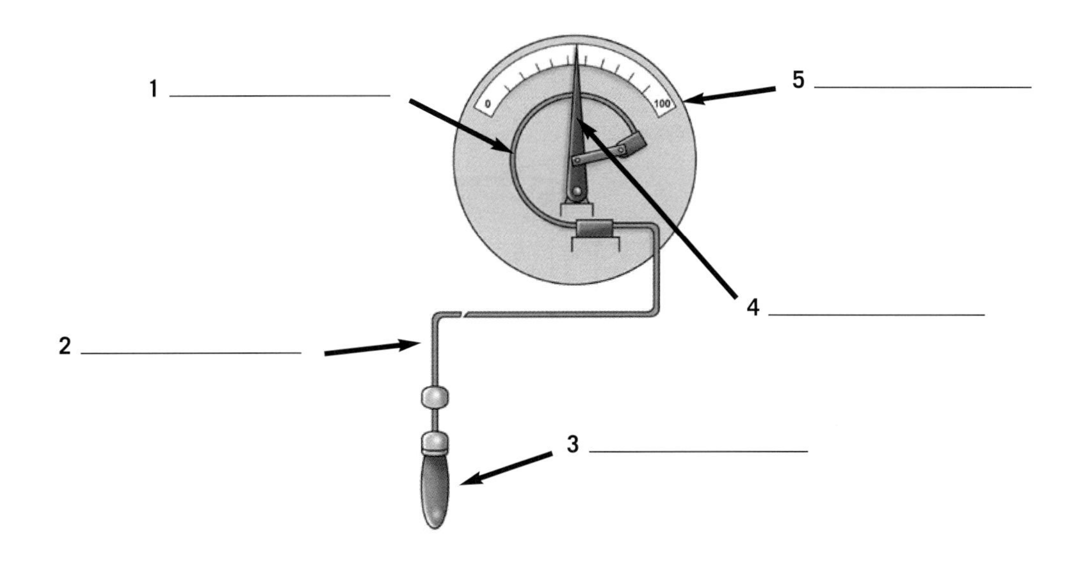

D Match the sentence halves below to make a complete text.

1 This is a	makes the pointer move.
2 It consists of	filled system thermometer.
3 This type of system	the temperature on the scale.
4 When the mercury in the bulb expands,	is completely filled with a liquid, usually mercury.
5 The spiral uncurls, and this movement	a bulb, a capillary tube, a bourdon tube, a pointer and a scale.
6 The pointer then indicates	it goes through the capillary tube and into the bourdon spiral.

E Listen and check your answers.

Lesson 7: Describing measuring devices (2): level

A Finish labelling the diagram with the words in the box.

pointer	tank	float	scale

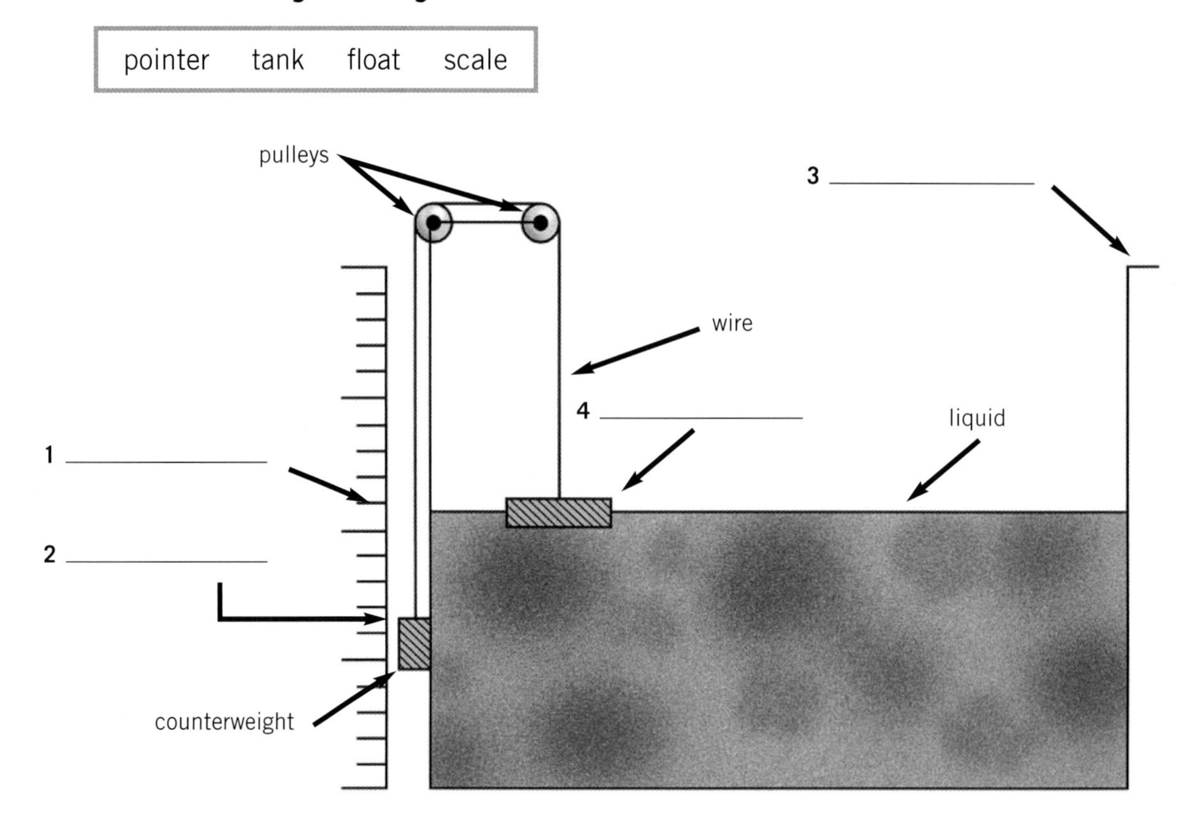

B Compare your answers with a partner. Discuss what you think the device measures, and what movements are involved.

C Listen to the description of the device.

D Complete the text with the correct form of the verbs, active or passive.

This diagram <u>shows</u> (*show*) a float system that <u>is used</u> (*use*) to measure the level of a liquid in a tank. The float [1] ______________ (*connect*) by a wire to a counterweight through a system of pulleys. A pointer on the counterweight [2] ______________ (*indicate*) the level on a scale. As the level of the liquid [3] ______________ (*decrease*), the float [4] ______________ (*get*) lower and the counterweight [5] ______________ (*pull*) higher. As the level of the liquid [6] ______________ (*increase*), the float [7] ______________ (*get*) higher and the counterweight [8] ______________ (*get*) lower. The change in level [9] ______________ (*show*) by the pointer.

E **Work with a partner. Look at the flow level meter (rotameter) below and discuss how it works.**

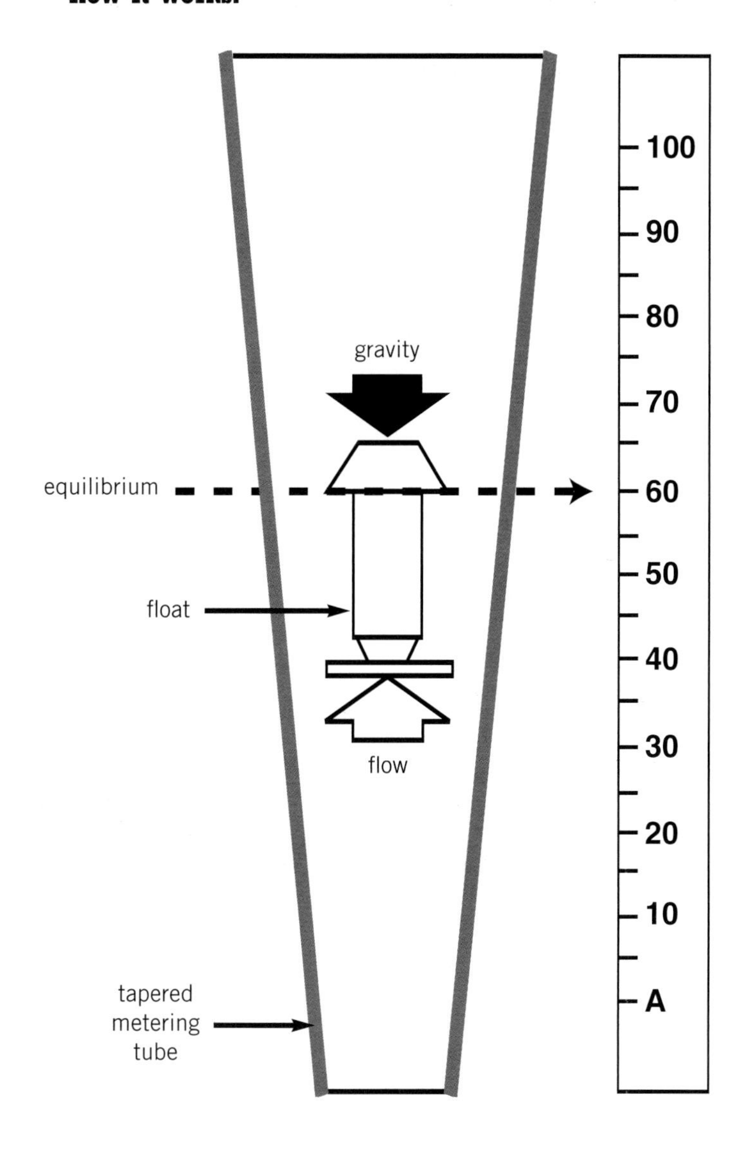

F **Write a description of the meter. Use some of the words in the box.**

float	rotate	rise	fall	flow	rate	top of the float	indicate

A rotameter consists of a glass tube with a measuring scale on it.

Lesson 8: Describing how tools work

A **Read the description of how a bench vice works. Finish labelling the diagram with the words in the box.**

fixed jaw	screw	body	bolt	slide

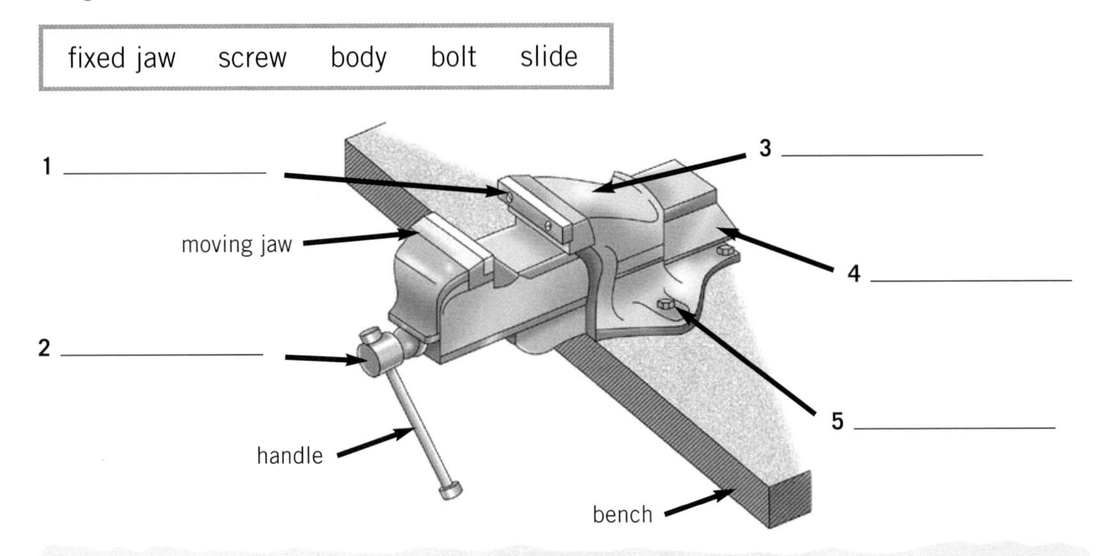

This is an engineer's bench vice. It can be used to hold your work. It is connected to the bench by bolts. It consists of a main body, with a fixed jaw to hold the work. A slide goes through the body. On the top of the slide there is a moving jaw. The handle on the left-hand side is attached to a screw which goes into the slide. When the handle is turned clockwise, the slide moves towards the bench, and the moving jaw holds the work against the fixed jaw. When the handle is turned anti-clockwise, the slide moves away from the bench and the moving jaw releases the work.

B **Listen to the description of how a bench drill works, then finish labelling the diagram with the words in the box.**

chuck
controls
motor housing
locking handle
base
operating lever

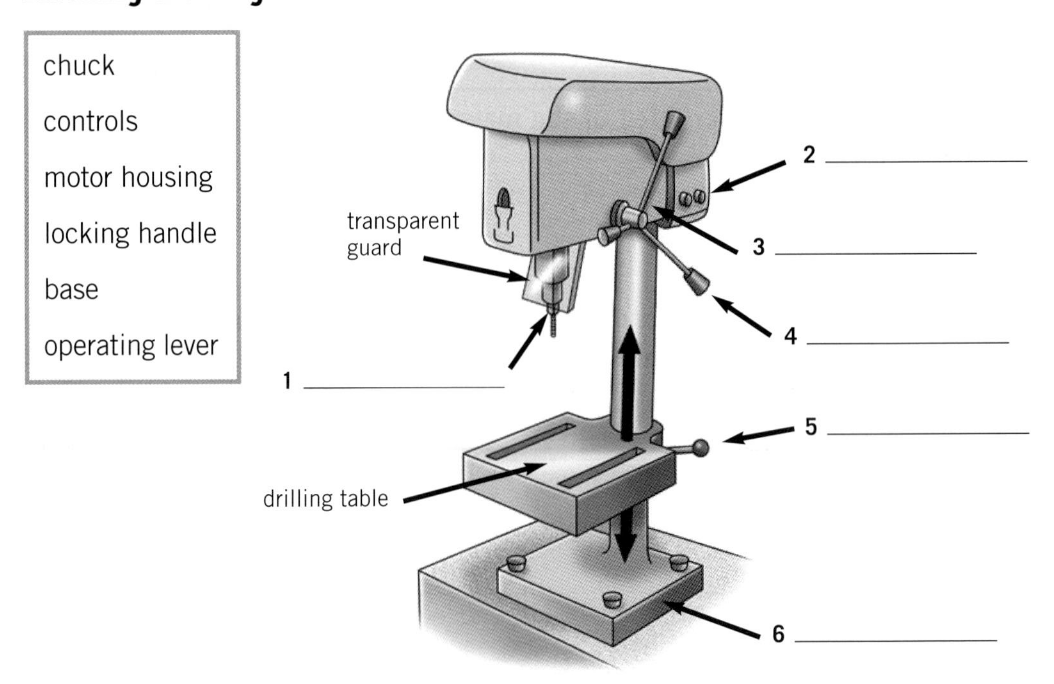

C Complete the description of a vernier caliper with the phrases in the box.

It consists of	is used to (x3)	hold	It can be used to	This is a

[1] ________________ vernier caliper. [2] ________________ measure internal or external dimensions. [3] ________________ a fixed jaw, a sliding head, a sliding jaw, a head lock, a clamp lock, a clamp and a fine-setting screw. The sliding head [4] ________________ measure internal dimensions. The sliding jaw [5] ________________ measure external dimensions. The locks and clamps [6] ________________ the sliding parts in place. The fine-setting screw [7] ________________ get exact measurements.

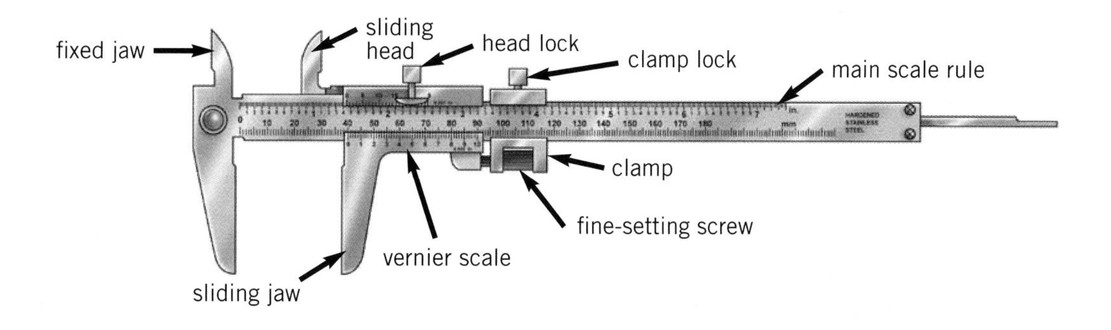

D Complete the description of the hacksaw below.

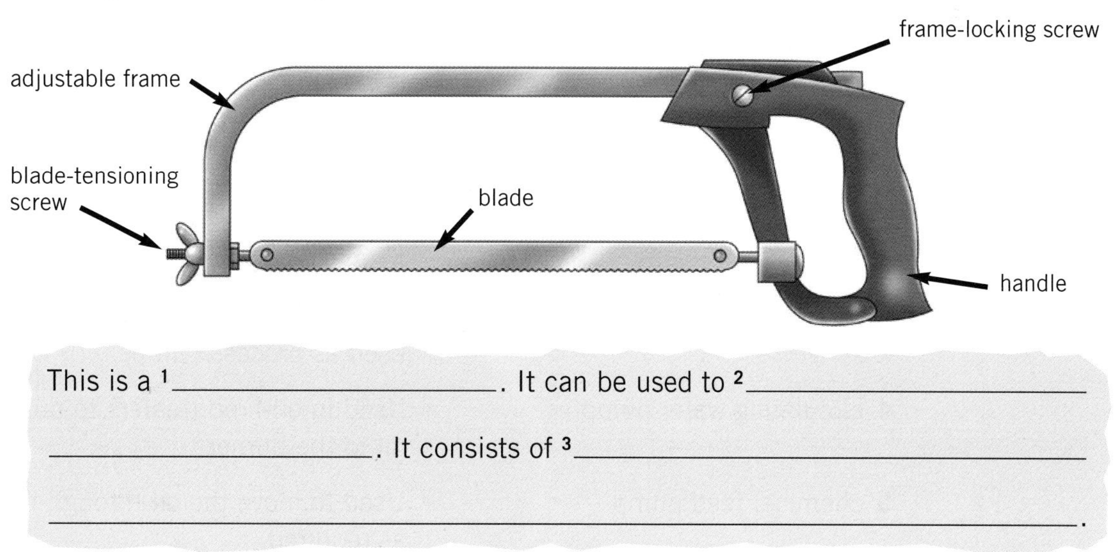

This is a [1] ____________________. It can be used to [2] ____________________
____________________. It consists of [3] ____________________
____.

E Listen and compare your description.

F Work with a partner. Choose one of the tools from the unit. Ask each other questions about how it works.

Lesson 9: Describing types of pumps

A Pumps are essential to the oil industry, and many different types of pump are used on site. Read the definitions below and decide which one is most appropriate.

1 A pump is a piece of equipment used to transfer liquid from one place to another.

2 A pump is a machine that gives energy to a liquid so that it moves.

3 A pump is a mechanism for moving gas, air or liquid from one place to another.

B Work in groups. Make a list of all the types of pump you see or use in everyday life.

____

____

____

____

____

C Match each of the more common types of pump found onsite with its usage.

1 main and auxiliary oil pumps	Used to supply water to plant fire lines.
2 fuel oil pump	Also called a cooling water pump. It is used to pump water through a heat exchanger such as a condenser or oil cooler.
3 lubricating oil pump	Small capacity units are used to pump chemicals into boilers; larger units are used as process pumps.
4 circulating water pump	Used in oil-fired treaters to pump fuel oil to the burners.
5 chemical feed pump	Used to move the oil through the pipeline as required.
6 fire pump	Used to circulate oil around a machine such as a turbine, engine, pump or compressor.

D Match the distances with the correct arrows, A, B and C.

1 Static suction lift

The pump has to lift the water through the suction line. The distance, measured vertically, that the intake of the pump is placed above the surface of the water is called the static suction lift. _____

2 Static discharge head

The vertical distance, in metres or in feet, from the centre line of the pump to the free surface of the water in the discharge tank is called the static discharge head. _____

3 Total static head

The vertical distance from the surface of the source of supply to the surface of the water in the discharge tank is called the total static head. This is the sum of static suction lift plus static discharge head. Thus, it is the total height the water is raised by the pump. _____

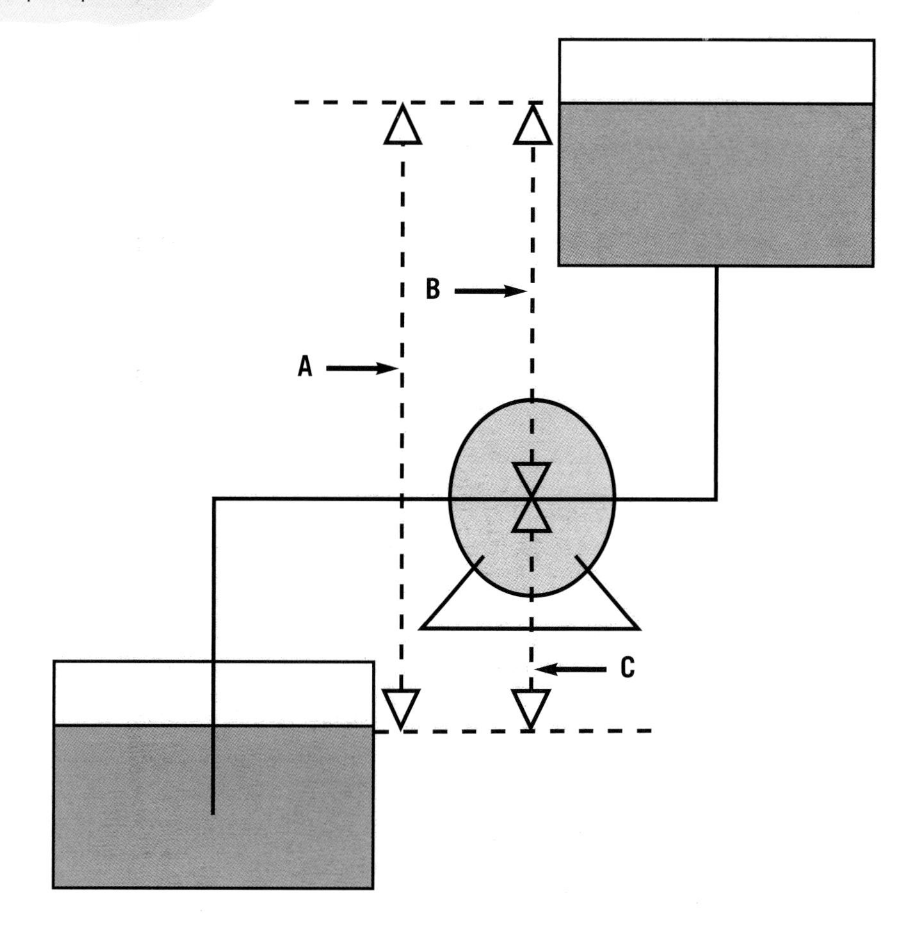

Lesson 10: Describing how pumps work

A **Discuss the pump system in the diagram. Label the different parts.**

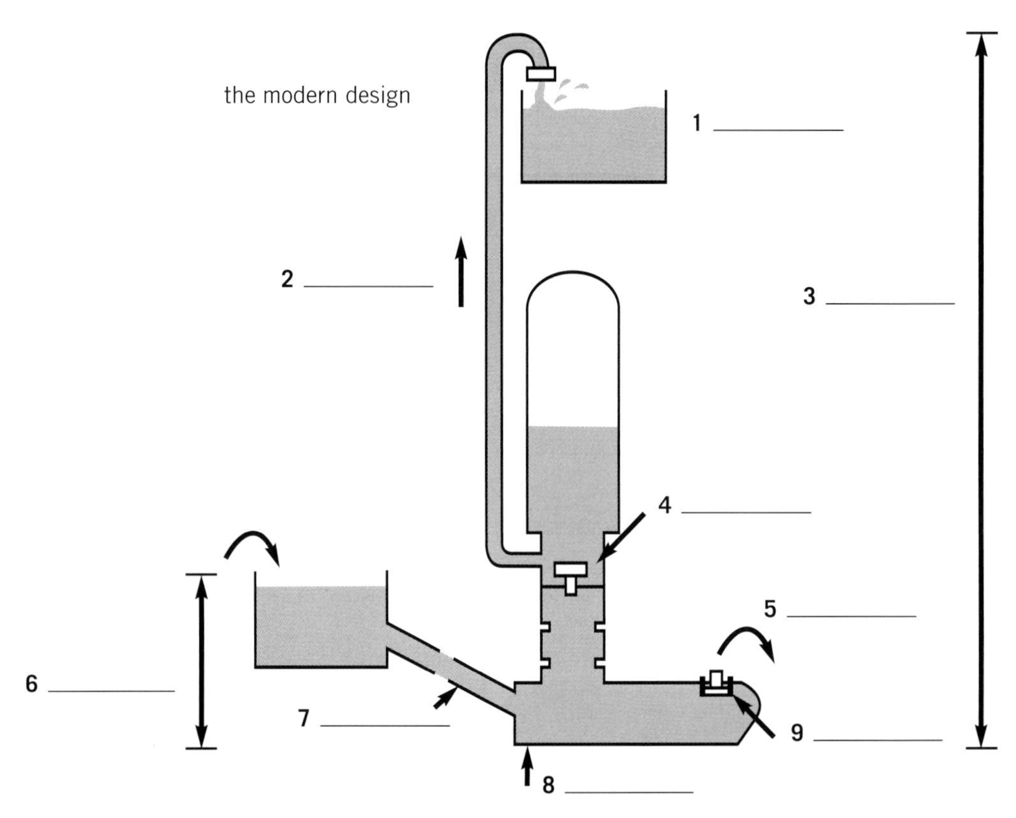

B **Choose the correct verb to complete each sentence. Discuss the difference in meaning between the two verbs.**

1 Domestic water pumps are used to *discharge* / *supply* water to kitchens and bathrooms.

2 Hot water is heated by the boiler, then *circulates* / *reverses* through the heating system.

3 Liquid is *drawn* / *injected* into the pump through a suction valve.

4 A piston *lifts* / *moves* in and out of the pump.

5 The piston *pulls* / *forces* water out into the pipe.

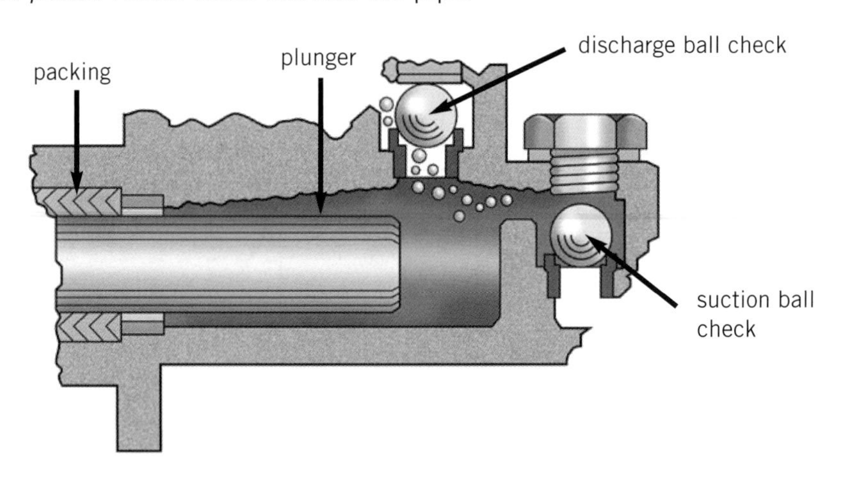

C **Read the information about a reciprocating pump and a single-acting pump. Complete the text with the verbs in the correct form, active or passive.**

Reciprocating pump

In this type of pump, the pumping action [1]______________ (*produce*) by the to-and-fro (reciprocating) movement of a piston or plunger within a cylinder. The liquid [2]______________ (*draw*) into the cylinder through one or more suction valves, then [3]______________ (*force*) out through one or more discharge valves by direct contact with the piston or plunger.

Single-acting pump

When the plunger [1]______________ (*move*) from right to left, the liquid [2]______________ (*draw*) into the cylinder through the suction ball check. When the plunger [3]______________ (*reverse*) and [4]______________ (*move*) from left to right, the liquid [5]______________ (*force*) out through the discharge ball check. The discharge ball check [6]______________ (*force*) open by the pressure of the liquid and, at the same time, the suction ball check [7]______________ (*force*) closed. The movement of the plunger in the cylinder in one direction [8]______________ (*call*) the stroke of the plunger. The distance the plunger [9]______________ (*move*) in and out of the cylinder [10]______________ (*call*) the length of the stroke. Only one side of the plunger [11]______________ (*take*) part in the pumping action, and water [12]______________ (*discharge*) only during one out of every two strokes. For these reasons, the pump [13]______________ (*call*) 'single-acting'.

D **Listen and check your answers.**

UNIT 3

Review: Describing equipment

A Work with a partner.
Student 1: Look at the picture of tools on page 217.
Student 2: Look at the picture of tools on page 227.

Take it in turns to tell your partner about your picture. Find six differences between the pictures without looking at your partner's picture.

B Work in two groups.
Group 1: Read the description of a valve on page 217.
Group 2: Read the description of a valve on page 227.

Answer the questions for the valve you read about.

1 What type of valve is this?

Valve A ______________________________

Valve B ______________________________

2 Does the valve have a handle?

Valve A ______________________________

Valve B ______________________________

3 What other parts does it have?

Valve A ______________________________

Valve B ______________________________

4 How does it open?

Valve A ______________________________

Valve B ______________________________

5 How does it close?

Valve A ______________________________

Valve B ______________________________

C Work with a partner who has information about a different valve. Ask questions about the other valve and write the answers.

D Ask your partner for a physical description of the valve they read about, then draw it in the box below. Label the diagram.

UNIT 3

Assess your skills: Describing equipment

Complete (✓) the tables to assess your skills.

I can ...	Difficult	Okay	Easy
• describe hand tools and identify their uses.			
• describe location using prepositions.			
• understand descriptions of tools and their parts.			
• understand descriptions of measuring devices.			
• listen to descriptions of devices and label diagrams.			
• describe how tools and devices work.			
• write descriptions of tools.			
• understand texts that describe how pumps work.			

I understand ...	Difficult	Okay	Easy
• the unit grammar: *can* to express ability and possibility			
the passive voice			
• the unit vocabulary (see the glossary)			

If there is anything you are not sure of, ask your trainer to revise the material.

UNIT 4

GIVING INSTRUCTIONS AND WARNINGS

The aims of this unit are to:

- provide the structures necessary to give and understand instructions
- describe ongoing situations/activities

By the end of this unit, you will be able to:

- identify different types of controls and explain their use
- describe basic operations
- identify dangerous situations and give warnings
- describe what is happening at the moment of speaking

Lesson 1: Following instructions

A Match each instruction with a diagram.

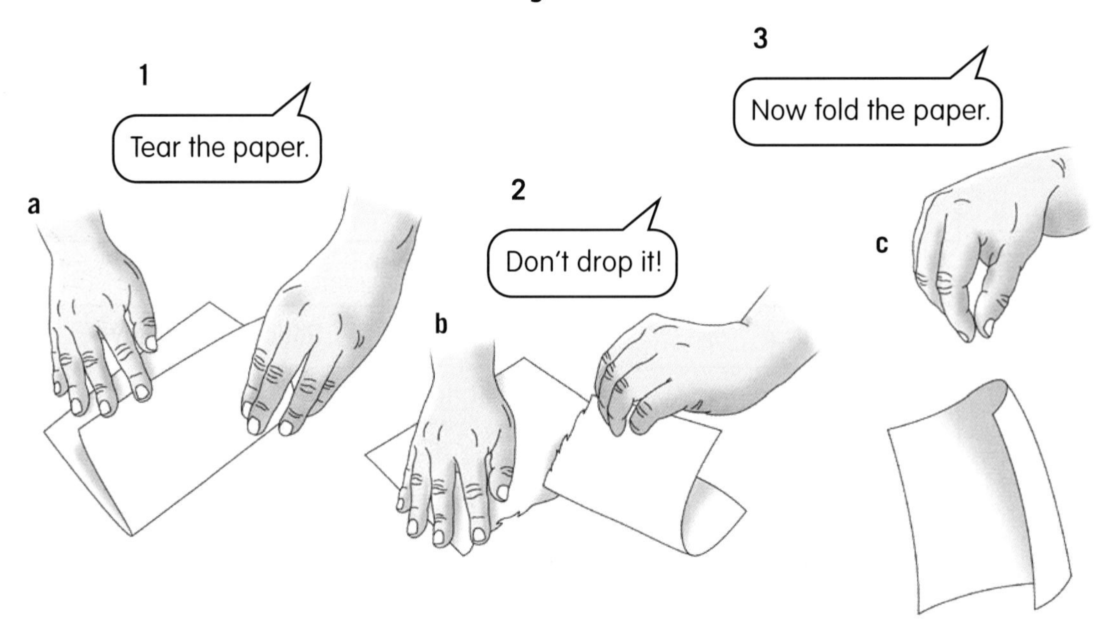

B We use the imperative to give instructions, warnings and commands. Complete the rule.

The imperative

Imperative sentences have no subject.
They are formed using the base form of the verb.
Negative imperative sentences are formed using ____________ + the base form of the verb.

C Rearrange the words to make instructions.

1 of paper / inches / inches by four / three / Cut / a piece / .

2 a horizontal line / one inch from / Draw / the top / across the paper / .

3 of the paper / into three / Draw two / vertical lines to / divide the bottom / equal parts / .

4 Carefully / vertical lines / tear along the / up to the horizontal / line / .

5 Fold the / towards you / the paper up / left section of / .

6 the right / the paper / section of / up away / from you / Fold / .

7 middle section / clip on the / Put the / paper / bottom of the / .

8 paper by / Hold the / and drop / your head / the top / above / it / !

D Listen and follow the instructions. What do you make?

E Read the instructions on how to send a letter. Complete each sentence with a suitable verb.

How to send a letter

1 First, ____________ your letter.

2 Then, ____________ it into three.

3 Next, ____________ the letter into an envelope.

4 ____________ the address on the front of the envelope.

5 After that, ____________ a stamp on the top right-hand corner of the envelope.

6 Finally, ____________ your letter to a postbox or post office and ____________ it.

F Work with a partner. Give each other instructions on how to do something. Write the instructions.

How to send an e-mail

1 First, ____.

2 Then, ____.

3 Next, ____.

4 After that, ____.

5 Finally, ____.

Lesson 2: Describing controls

A Complete the instructions with verbs from the box. (Sometimes more than one verb is possible.)

turn (on/off) push pull press release

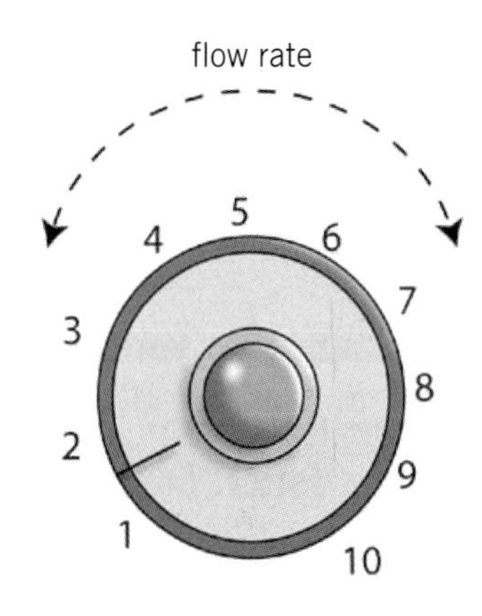

1 Turn __________ the dial.

2 __________ the button.

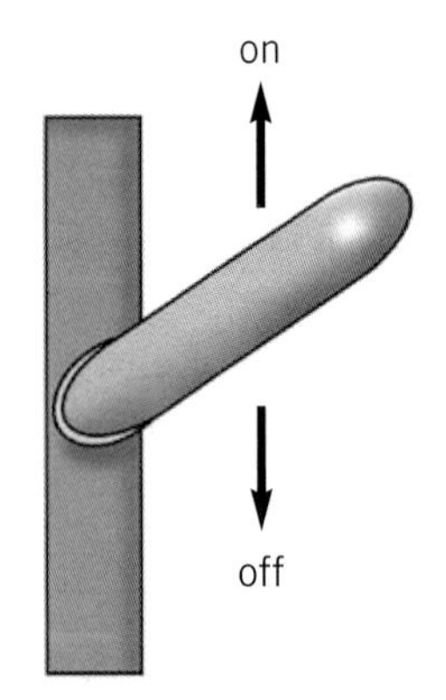

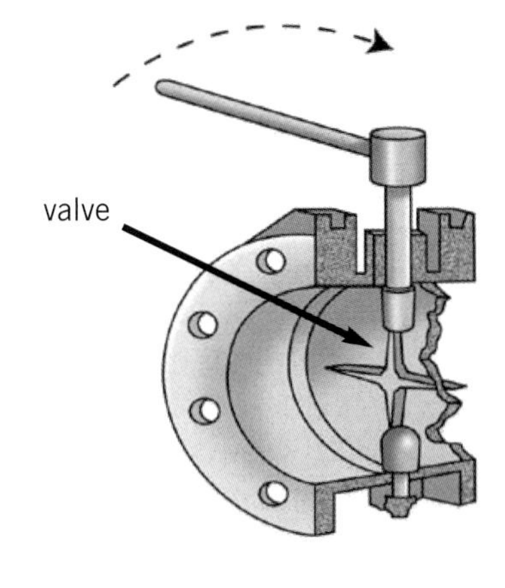

3 __________ the lever.

4 __________ the handle.

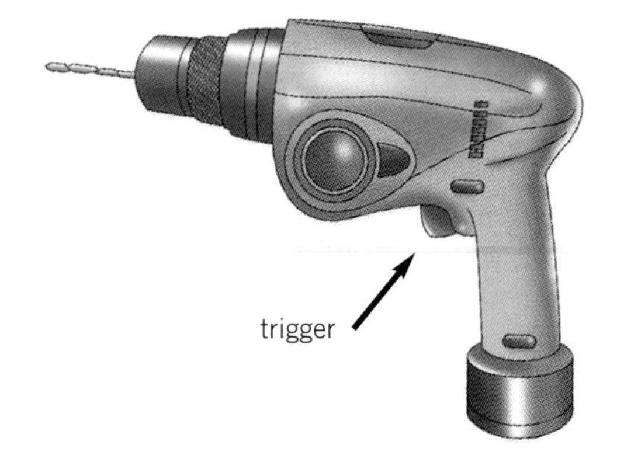

5 __________ the switch.

6 __________ the trigger.

B **Look at the pictures on page 70. Discuss them with a partner.**

C **Complete the sentences. Match them with the pictures.**

1 Push the lever up to turn the machine on, and push it down to turn the machine off.

2 ________ the trigger to start the drill, and ________ the trigger to ________ the drill.

3 Turn the ________ clockwise to increase the flow rate, and ________ it ________ to ________ the flow rate.

4 ________ the button on the ________ to start the machine, and ________ the button on the ________ to ________ the machine.

5 Press the top ________ to move up, and ________ the ________ button to move ________.

6 ________ the handle ________ to ________ the valve, and ________ it ________ to ________ the valve.

D **Listen and check your answers.**

E **Complete the instructions on using a car.**

1 ________________________ to lock the car.

2 ________________________ to unlock the car door.

3 ________ the steering wheel ____________ to turn left.

4 ________ the foot brake ____________ to stop the car.

5 ____________ the accelerator ____________ to ____________ your speed.

Lesson 3: Giving instructions for using tools

A Bob wants to drill a hole. He asks Vasily for instructions. Listen and put the instructions in the right order (A–M).

1 Mark the hole. ☐
2 Tighten the chuck. ☐
3 Remove the bit from the hole. ☐
4 Start the drill. ☐
5 Stop the drill. ☐
6 Drill the hole. ☐
7 Disconnect the drill from the power. ☐
8 Connect the drill to the power. ☐
9 Place the bit over the mark. ☐
10 Remove the drill bit. ☐
11 Measure the work. ☐
12 Loosen the chuck. ☐
13 Attach the drill bit. ☐

B **Work with a partner. Decide (✓) which sentences give sensible instructions for someone in a workshop.**

1 Do not eat. ☐

2 Remove jewellery such as rings. ☐

3 Avoid running. ☐

4 Adjust your hair so that it is not tied back. ☐

5 Always wear safety goggles. ☐

6 Loosen your shoelaces. ☐

7 Check that you are wearing the correct PPE. ☐

8 Take off your shoes. ☐

C **Work in pairs. Give each other instructions.**
Student 1: Check that your partner's instructions are correct. Look at page 218.
Student 2: Check that your partner's instructions are correct. Look at page 228.

Student 1
You want to …
1 change a light bulb.
2 cut a piece of pipe.

Student 2
You want to …
1 change a fuse in a plug.
2 climb up a ladder.

D **Choose one of the activities from exercise C. Write four sentences giving instructions on how to do it (do not look at the models in the book). Then compare your sentences with the ones on pages 218 and 228.**

1 ______________________________

____

2 ______________________________

____

3 ______________________________

____

4 ______________________________

____

Lesson 4: Describing and explaining things that are happening now

A **Listen to the conversation between Bob and Ahmed. Decide whether the statements below are true (T) or false (F).**

1 Bob always works late. ☐

2 There is a problem with a piece of equipment. ☐

3 Bob needs Ahmed's help. ☐

4 Bob is repairing the system. ☐

5 The system is too hot. ☐

6 The engineer is coming later tonight. ☐

B **Complete the table below.**

The present continuous tense is formed using:

Subject + *be*	Verb + *-ing*	
I'm (I am)	________	late tonight.
The valve ______	working	properly.
Bob and Mustafa ______	repairing	the system.
What ______ you	________	now?

C **Look at the picture and describe what is happening. Use the verbs in the box and the present continuous.**

hold	wear	point	shout

The manager is pointing at John's's ear protectors.

D **Work with a partner.**
Student 1: Think of different hand tools and mime using them.
Student 2: Guess what your partner is doing.

E **Read the following exchanges, then complete the sentences.**

1 **A:** Why ______________________________? (disconnect the power)

B: To change the fuse.

2 **A:** Why ______________________________? (press that button)

B: ______________________________

3 **A:** Why ______________________________? (increase the volume)

B: ______________________________

F **Practise your dialogues with a partner.**

Lesson 5: Giving warnings

A Look at these imperative sentences. Which are instructions (I) and which are warnings (W)?

1 Turn it off! ☐ ☐

2 Turn the dial carefully. ☐ ☐

3 Don't go in there! ☐ ☐

4 Slow down! ☐ ☐

5 You can't smoke in here! ☐ ☐

6 Be quiet and listen! ☐ ☐

7 Be careful! ☐ ☐

8 First, turn the machine on. ☐ ☐

B Now listen and tick the sentences you hear.

C Complete the sentences using suitable verbs and tenses. Then write a warning.

Situation	Warning
1 Somebody <u>is using</u> (*use*) a circular saw without safety goggles.	*Put some safety goggles on!*
2 Somebody ____________ (*talk*) when the supervisor is explaining a job.	____________!
3 Somebody ____________ (*enter*) a restricted area.	____________!
4 Somebody ____________ (*carry*) a cup of coffee and ____________ (*not look*) where they are going.	____________. *Watch where you're going!*
5 Somebody ____________ (*smoke*) near some flammable containers.	____________!

D What warnings do you give in the following situations?

1 Somebody is walking in the workshop with no PPE.

____

2 Somebody is using scaffolding without a safety harness.

____

3 Somebody is driving over the speed limit.

____

4 Somebody is using dangerous chemicals.

____

5 Somebody is standing close to some dangerous equipment.

____

E Work with a partner.
Student 1: Look at page 218.
Student 2: Look at page 228.

Tell your partner what is happening in your pictures. They should respond with an appropriate warning.

F Correct the sentences below.

1 No smoke in the workshop!

____

____

2 Why you are not wearing safety boots?

____

____

3 I need a spanner for tighten this nut.

____

____

4 Don't forget to turning the computer off.

____

____

5 Pick that box up careful.

____

____

6 This mobile phone not working properly.

____

____

Lesson 6: Comparing temporary and permanent situations

A **Read the text and underline the verbs that are used in the present simple. Circle the verbs that are used in the present continuous.**

Alan is a technical trainer. He comes from Scotland, but at the moment he's living in Azerbaijan. He works at the training centre every day, and enjoys meeting people from other cultures. Today is a holiday, so he is having coffee with a friend and practising his Azeri.

B **Read the rules and add examples from the text.**

Present simple

We use the present simple to describe a general situation that is always true:

- for facts. Example: ______________________________
- for habits. Example: ______________________________
- with certain verbs such as *be*, and with verbs of opinion or perception, e.g., *love, hate, see, hear*. Example: ______________________________

Past — Now — Future

Present continuous

We use the present continuous to describe a specific situation that is happening around now:

- for actions in progress. Example: ______________________
- for temporary situations. Example: ______________________

Past Now Future

C **Match the correct halves of the exchanges.**

1 The room gets too hot.	Open the window.
2 The room is getting too hot.	Install air conditioning.
3 The valve doesn't work.	Yes, but he's on holiday at the moment.
4 The valve isn't working.	No, we never use it.
5 Does he work on the rig?	Check it.
6 Is he working on the rig?	Not now, but I'll need it later.
7 Are you using this torch?	No, I've got a better one.
8 Do you use this torch?	Yes, he'll be back next week.

D **Complete the paragraph about Hassan with the verbs in the box. Use the present simple or the present continuous.**

be	know	find	explain	have	do (x2)	like	study	show

Hassan [1]______________ to be a production engineer. He [2]______________ a degree, and at the moment he [3]______________ a training course. The course involves studying in the workshop and working on site. He [4]______________ the training that happens on site because it [5]______________ how things work in action and not just in theory. This week they [6]______________ pigging. He [7]______________ some things already, but [8]______________ the course useful, especially when their trainer [9]______________ about the different types of pig, and what they [10]______________.

E **Listen and check your answers.**

Lesson 7: Talking about problems in the workshop

A **Look at the pictures. Vasily is doing some jobs in the workshop. What is he doing right? What is he doing wrong? Write sentences using the present continuous.**

Right

1 *He's wearing his hair tied back.* ____________

2 ________________________

3 ________________________

Wrong

1 *He isn't wearing ...* ____________

2 ________________________

3 ________________________

4 ________________________

5 ________________________

B **Work in groups. Write some important rules for working in the workshop. Use the imperative.**

C **Listen. What are the problems?**

D **Listen again and complete the dialogue. Put the verbs into their correct form.**

Vasily: What [1]______________ (*be*) the problem?

Alan: The switch [2]______________ (*not/work*), so the machine [3]______________ (*not/stop*).

Vasily: [4]______________ (*try*) the emergency shut down.

Alan: That [5]______________ (*not/work*) either.

Vasily: [6]______________ (*turn off*) the power and [7]______________ (*restart*) in ten minutes.

E **Work with a partner. Think of possible solutions for the problems below. Write your ideas in the boxes.**

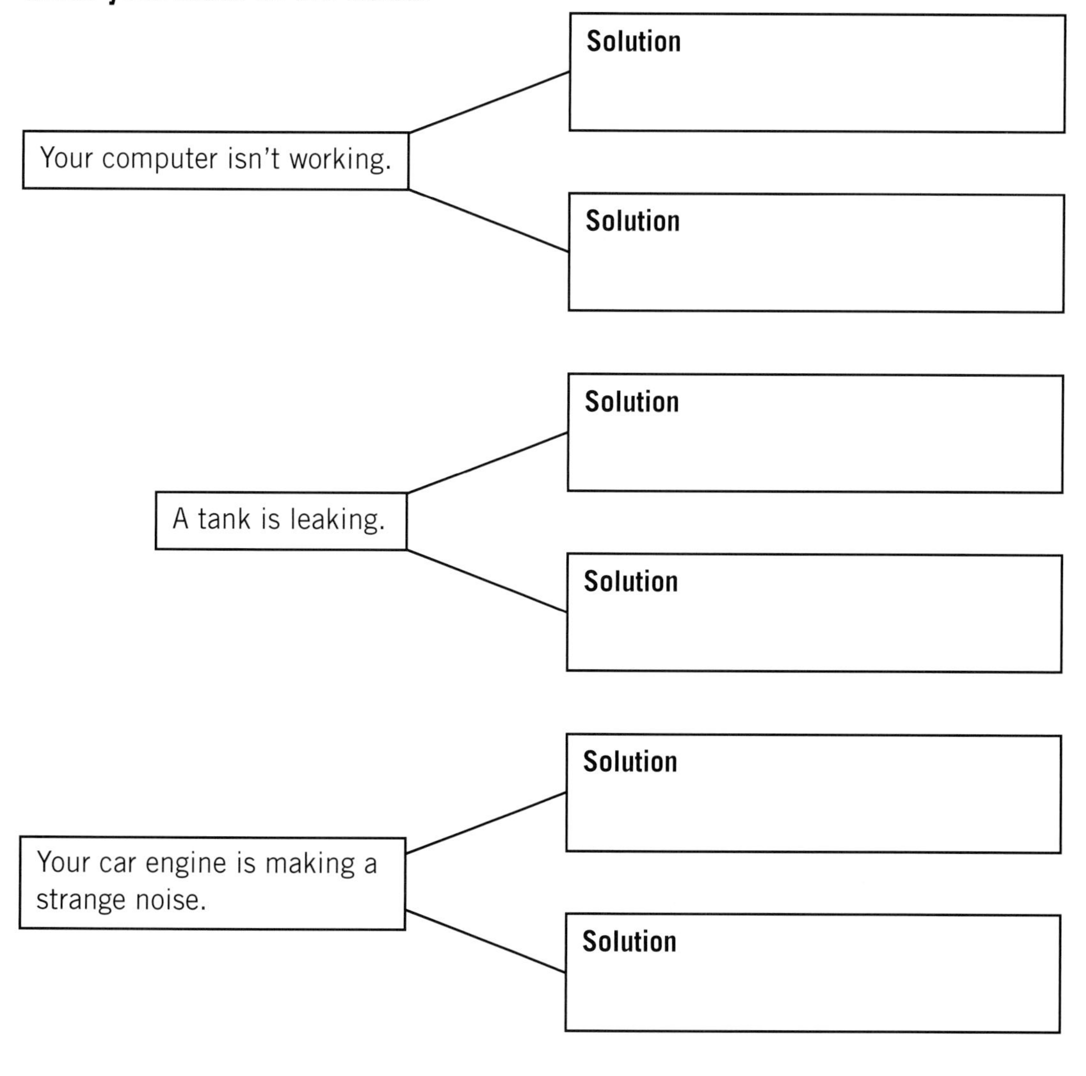

F **Practise similar dialogues to the one in exercise D. Use the situations above.**

Lesson 8: Talking about the weather

The weather

We can talk about the weather in two ways:

Present continuous	or	***It is* + adjective**
It's raining.		*It's rainy.*
The sun's shining.		*It's sunny.*

A **Identify the weather conditions in the pictures below. Which weather conditions can we only talk about in one way?**

wind	rain	snow	sun	cloud	fog

1

2

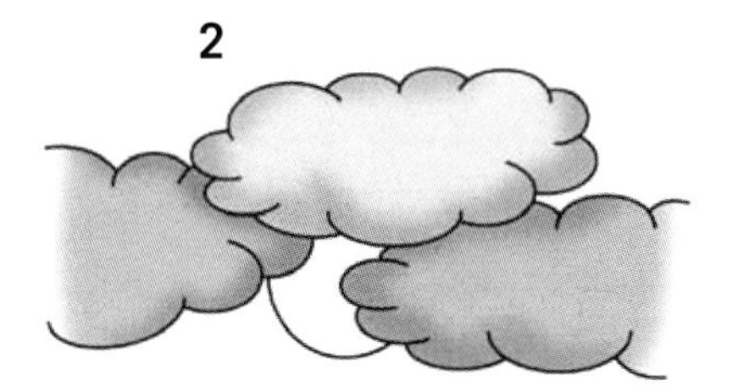

3

4

5

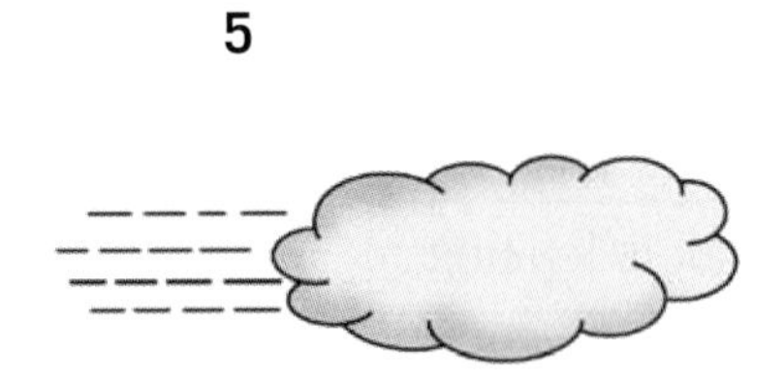

6

B **Match each weather condition with its description.**

1 ice	electricity discharges from a cloud
2 mist	the sound of electricity discharged from a cloud
3 lightning	ice falling from clouds
4 thunder	rain and snow together
5 sleet	thin fog
6 hail	frozen water on the ground

C Listen to the weather forecast and answer the questions.

1 What was the weather like in Scotland yesterday?

____

2 How is the weather better today?

____

3 Where is it snowing?

____

4 What could cause a problem?

____

D Complete the sentences.

1 There was __________ and __________ in many parts of Scotland yesterday.

2 At the moment it's __________ in Aberdeen.

3 Most areas are quite __________ today.

4 __________ over the hills will clear later.

5 __________ __________ will make it feel cold.

6 Watch out for __________ __________.

E Work with a partner. Try to explain the weather shown in the icons below.

1 

2 

3 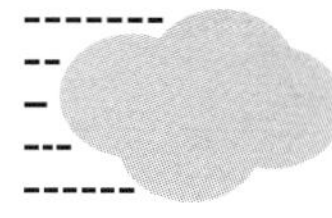

4 

5 

6 

7 

8 

9 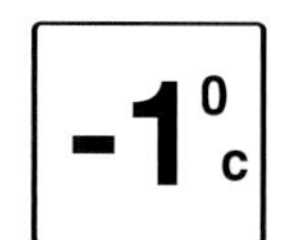

F Work with a partner. Discuss these questions, then present your ideas to the group.

1 What effects do weather conditions have on oil industry operations?

2 Which conditions are dangerous for the following operations? Give warnings.

a helicopter flights **b** crane operations **c** digging **d** drilling

Lesson 9: Talking about crane controls

A Read the names of the controls in the box.

1 Which are used in a car?

2 Which are used in a crane?

3 Which are used in both?

steering wheel	wheels	accelerator	boom
brake	jib	emergency stop button	lever

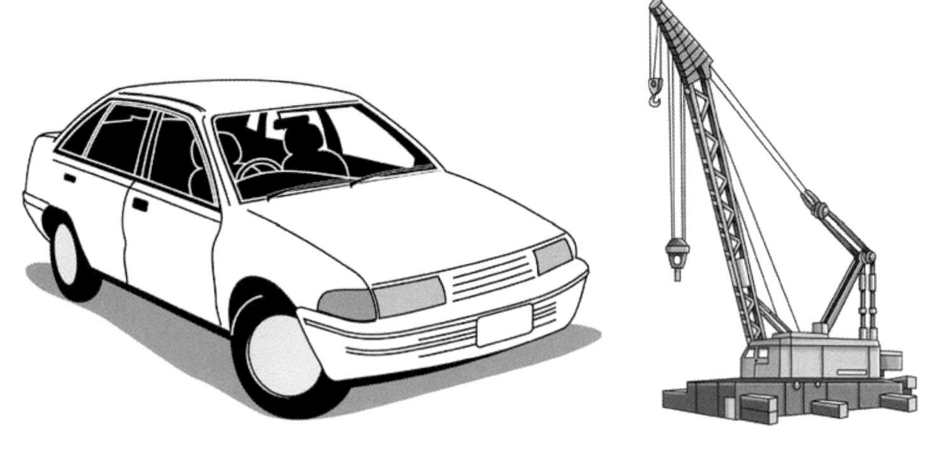

B Read the text below about crane controls, then answer the questions.

A crane has a lot of different controls. Some control the movement of the crane, some control the movement of the jib. The accelerator, brake and steering wheel control the movement of the crane. The accelerator increases the speed of the crane, the brake stops the movement of the crane, and the steering wheel controls the direction of the crane. Three levers and one button control the movement of the jib. Lever A controls vertical movement of the jib, lever B controls horizontal movement of the jib, lever C controls the vertical movement of the load and button A is an emergency stop.

1 Do the same controls control the crane and the boom?

____

2 What makes the crane go faster?

____

3 What does the steering wheel do?

____

4 Can lever A move the boom to the left and right?

____

5 Can lever C raise the load?

____

6 What do you do if there is an incident?

____

C Label the diagram of the interior of a crane cab with the words from the box.

boom	jib	load	steering wheel	lever	button	accelerator	brake

Lesson 10: Talking about instructions for crane operations

A Look at the groups of words. Find one word that has a different meaning to the others in each row.

1 raise	hoist	lift	load
2 signal	slew	swing	slide
3 extend	telescope	retract	lengthen
4 stop	cease	inch	halt

B Match each group of words above with a diagram.

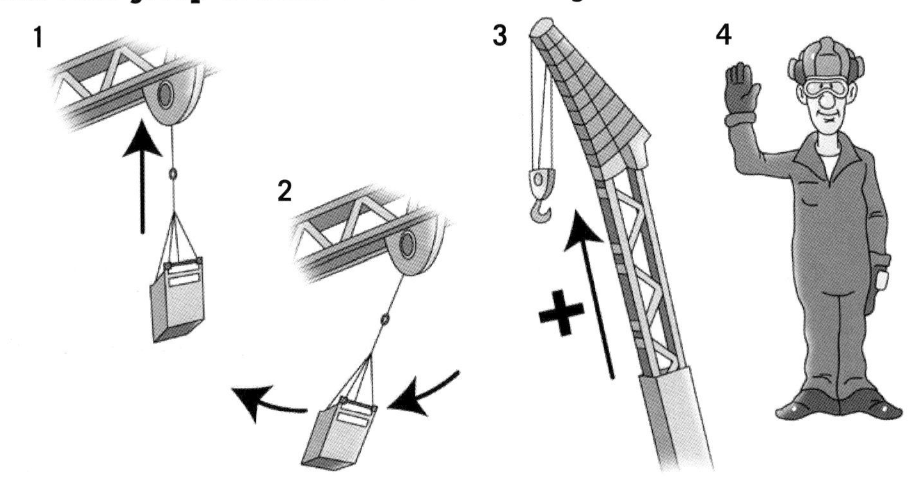

C Complete the text about crane operations with the words and phrases in the box.

vision	load	careless	clear and correct	swing	signals	accident
crane operators	at all times	stop				

If [1] ____________________ are unable to see what they are doing clearly, they may cause a lot of damage. For example, they may attempt a lift before the [2] ____________________ is secured, they may lower the load on top of people or they may [3] ____________________ the boom into the structure.

The signal person, or spotter, is the person who helps the operator use the crane safely by signalling what to do. Three things are important. The operator must see the signal person, or be in contact [4] ____________________, the signals they give the operator must be [5] ____________________, and both must understand the meaning of each signal.

Problems that can happen

As the boom swings, the operator can lose sight of the signal person if they don't stay within the operator's field of [6] ____________________. Poor signals are difficult to understand. If the spotter is [7] ____________________ or unclear about the signals given to the operator, the operator will be unsure what to do, or worse, may mistake the signal for another one, causing an [8] ____________________.

Crane safety tips

- Have one signal person for each crane.
- If you're the operator, know who and where your signal person is at all times. If you're not sure what is happening, [9] ____________________ and wait for a signal.
- Both operators and signal persons must know each crane signal. If you are unsure, take the time to learn the [10] ____________________ before the next lift.

D Answer the questions.

1 What is the job of the operator?

2 What is the job of the spotter?

3 When is it necessary to see the spotter?

4 What can happen if the spotter does not give clear instructions?

E Work with a partner. Look at the guidelines below for hand signals used for crane operations.

Student 1: Make the signals.
Student 2: Write down what your partner is signalling.

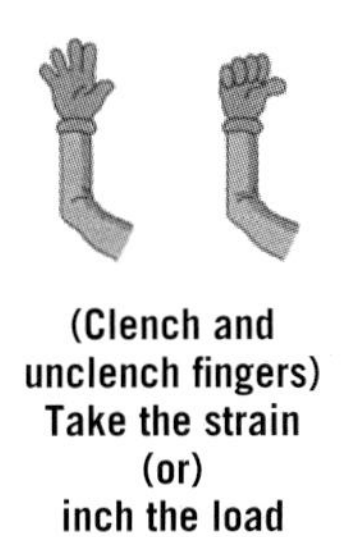

(Clench and unclench fingers) Take the strain (or) inch the load

Signal with one hand, other hand on head

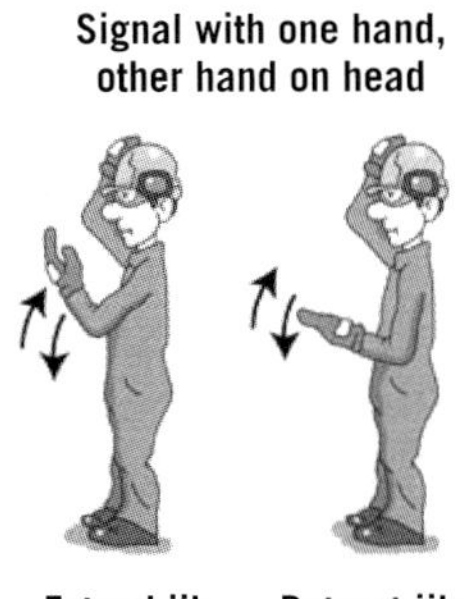

Extend jib Retract jib (Telescoping jib)

Operations cease

Travel to me Travel from me (Signal with both hands)

Slew in direction indicated

Signal with one hand, other hand on head

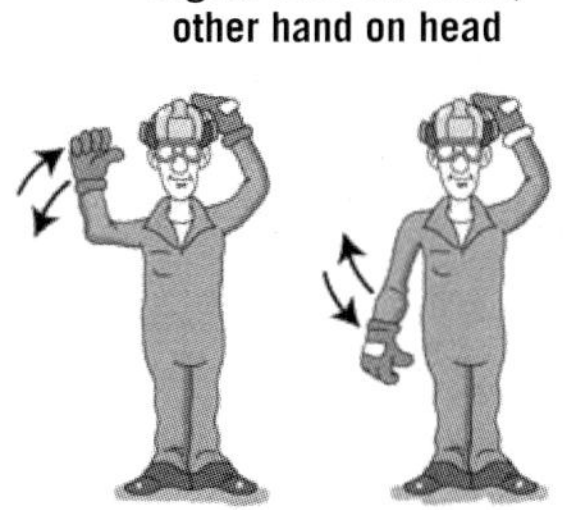

Jib up Jib down (Derricking jib)

Stop Emergency stop

Hoist Lower

UNIT 4 Review: Giving instructions and warnings

A Find ten words in the wordsearch. Five words are connected with the weather and five words are connected with cranes.

O	Z	S	J	I	B	E	R	S	U	D	L
U	E	X	T	H	N	G	U	P	S	H	I
G	X	A	I	O	S	B	T	O	H	A	G
T	P	L	L	I	T	N	W	T	J	T	H
A	C	C	E	L	E	R	A	T	O	R	T
B	O	S	V	I	E	W	B	E	N	A	I
O	S	H	E	N	R	N	T	R	M	I	N
M	I	A	R	X	I	D	C	R	A	N	G
I	O	L	O	E	N	I	N	D	B	I	C
S	N	L	N	D	G	O	N	I	T	N	R
T	B	O	I	B	W	E	F	O	G	G	Y
O	D	W	T	B	H	Z	C	I	M	E	A
T	H	U	N	D	E	R	A	C	T	E	R
A	F	I	R	D	E	H	L	L	Y	T	N
N	S	H	E	X	L	G	O	N	A	S	D

B Complete the sentences with words from the wordsearch.

1 When you drive a car or a crane, you hold the ________________.

2 If you want the crane to move faster, press the ________________.

3 The first ________________ controls the vertical movement of the ________________.

4 The ________________ gives signals to the crane operator.

5 It was stormy last night. We heard the ________________ and saw the ________________.

6 Visibility is quite good. There's a slight ________________ over the sea, but it isn't ________________.

C Work with a partner.
Student 1: Look at page 219.
Student 2: Look at page 229.

Take it in turns to describe the picture without letting your partner see it. Your partner should say what they think is happening in your picture.

UNIT 4

Assess your skills: Giving instructions and warnings

Complete (✓) the tables to assess your skills.

I can ...	Difficult	Okay	Easy
• follow instructions.			
• give instructions clearly.			
• identify controls and explain their use.			
• understand and give warnings.			
• talk about permanent and temporary situations.			
• talk about the weather and its effect on operations.			
• write descriptions of tools.			
• understand texts about crane operations and signals.			

I understand ...	Difficult	Okay	Easy
• the unit grammar: the imperative			
present simple			
present continuous			
• the unit vocabulary (see the glossary)			

If there is anything you are not sure of, ask your trainer to revise the material.

UNIT 5 DESCRIBING SYSTEMS

The aims of this unit are to:

- provide the structures necessary to describe systems
- explain technical drawings

By the end of this unit, you will be able to:

- identify the parts of a system
- assess the advantages and disadvantages of systems
- explain the consequences of actions
- identify some process and instrument symbols
- construct logical system descriptions

Lesson 1: Describing systems and devices

A **Read the definitions and write what they are describing.**

1 The movement or flow of electric charge: **c** u r r e n t

2 A machine that changes mechanical energy into electrical energy: **g** __ __ __ __ __ __ __ __

3 A route that an electric current follows: **c** __ __ __ __ __ __

4 A safety device that protects an electric circuit from too much current: **f** __ __ __

5 A device that measures and records amount or volume, e.g., the flow of gas, oil or electricity: **m** __ __ __ __

6 A device that accepts video signals from a computer and displays information on a screen: **v** __ __ __ __ __ **d** __ __ __ __ __ __ **u** __ __ __

7 A device that warns of danger by making a sound or giving a signal: **a** __ __ __ __

8 A device that controls or regulates the flow of liquid or gas in pipes: **v** __ __ __ __

B **Which devices are these? Write the names.**

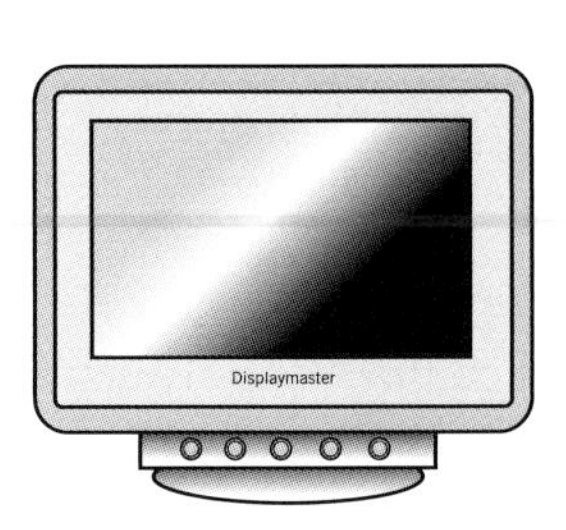

1 ____________ 2 ____________ 3 ____________ 4 ____________

Systems

To describe a system, you must identify the parts of the system, the purpose of those parts and how they interact with each other. Descriptions must be clear and concise so people can understand the system easily.

C Read the following system description and decide whether the description is clear.

Pressure measurement systems

A typical pressure measurement system consists of a monitoring device, a generator, a back-up generator, a relay, an alarm, a control panel and a visual display unit. The monitoring device measures the pressure in the pipeline and takes the information, via the relay, to the visual display unit. If the pressure reaches a preset low level, the alarm sounds and alerts the operator. The operator can then use the control panel to initiate an emergency shutdown. Power is provided to the system by a generator. There is a back-up generator which comes online in the event of a failure in the primary generator, or when the primary generator is undergoing maintenance.

D Answer the questions. Check your answers with a partner.

1 How many components are there in the system? What are they?

____

2 What does the monitoring device do?

____

3 What happens if there is a problem with low pressure?

____

4 What does the control panel do?

____

5 When does the back-up generator start?

____

6 What is the purpose of the system?

____

____

____

____

Lesson 2: Describing heating systems

A **Look at the heating system below and read the description of how it works. Complete the text.**

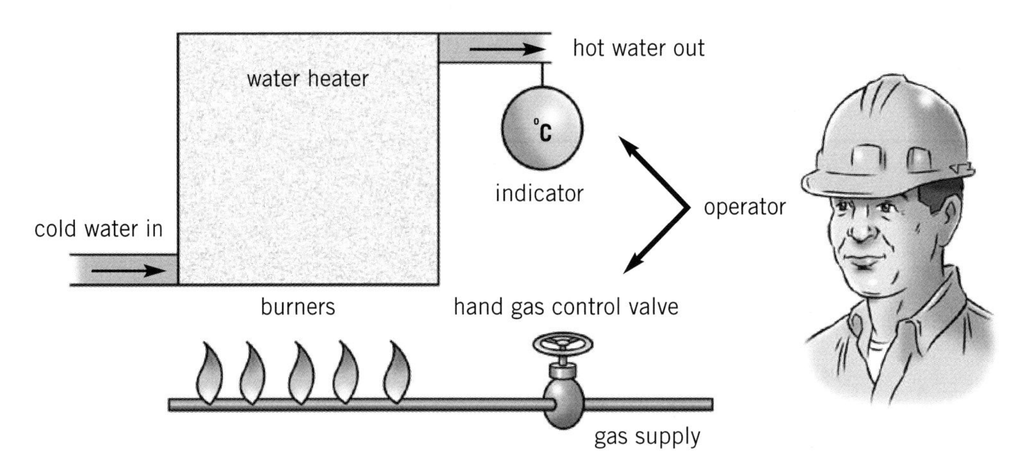

Open-loop systems

Open-loop systems are manual control systems. The system is controlled by an operator. The system includes a water heater tank, an 1__________, burners and a 2__________ __________ __________ __________. 3__________ __________ enters the tank through the 'in' pipe at the bottom. The water is heated by 4__________ below the tank. 5__________ __________ rises to the top of the tank, and leaves it by the 6'__________' pipe. An indicator shows the 7__________ of the water in the 'out' pipe. The temperature can be adjusted by the operator using the hand gas control valve to change the 8__________ supply to the 9__________.

B **Try to complete the description of the closed-loop system.**

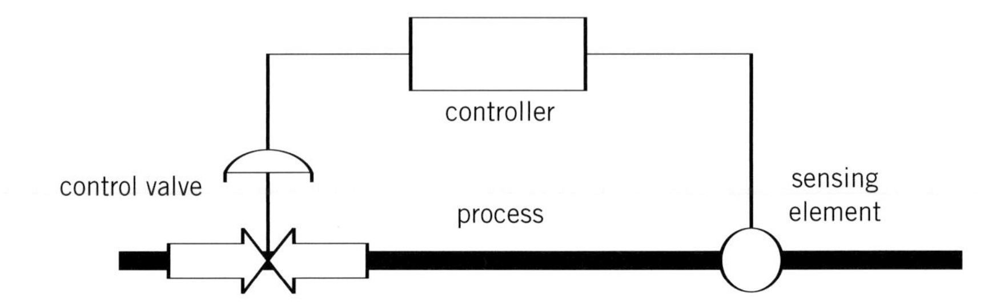

Closed-loop systems

Closed-loop systems are automatic control systems; a controller takes the place of the operator. The system consists of [1] ____

____. The sensing element monitors

[2] ____.

The flow can be adjusted by the controller sending [3] ______________________________

____, which regulates the flow.

C **Listen and check your answers.**

D **Work with a partner.**
Student 1: Look at page 219. Try to think of disadvantages of open-loop systems and advantages of closed-loop systems.
Student 2: Look at page 229. Try to think of advantages of open-loop systems and disadvantages of closed-loop systems.

Compare your ideas.

Lesson 3: Describing alarm systems

A **What is the difference between the two alarms in the pictures?**

B **Listen to someone describing alarm systems. Decide whether the statements below are true (T) or false (F).**

1 Alarms always involve sound. ☐

2 Alarms have different types of switches. ☐

3 Alarms are activated when processes operate outside their usual range. ☐

C **Listen again and complete the notes.**

Types of alarm

- ______________________: warns operator that there is a problem.
- ______________________: shows operator where the problem is.

How an alarm works

- Basic ON/OFF ______________________ that uses a limit-sensing device ______________________________________.
- When the machine or equipment operates outside the normal operating range, e.g., ______________________________________, the alarm goes off.

Types of limit-sensing device

- ____________, ____________, float-operated or flow-actuated switch.

D **Match each consequence in the box below with an action in the table.**

the alarm sounds.	the oil flows to the overflow tank.	the lights are red.	the back-up system comes online.

Action	Consequence
If there is a fire in the building,	
If the primary system fails,	
If the switch is in the OFF position,	
If the valve is closed,	

Conditional sentences

A conditional sentence uses *if*. It explains the consequence of an action.

Action	**Consequence**
If the pressure is high,	*the alarm sounds.*

Zero conditional

A zero conditional sentence describes facts or things that are generally true. We often use a zero conditional sentence to describe systems.

Action	**Consequence**
If + present simple,	present simple

E **Complete the sentences below using the same pattern.**

1 If the alarm goes off, we ______________________________.

2 If ______________________________, the alarm goes off.

3 If there is an accident, ______________________________.

4 If ______________________, ______________________.

Lesson 4: Describing how electrical systems work

A Work with a partner. Look at the two diagrams. Discuss how they work.

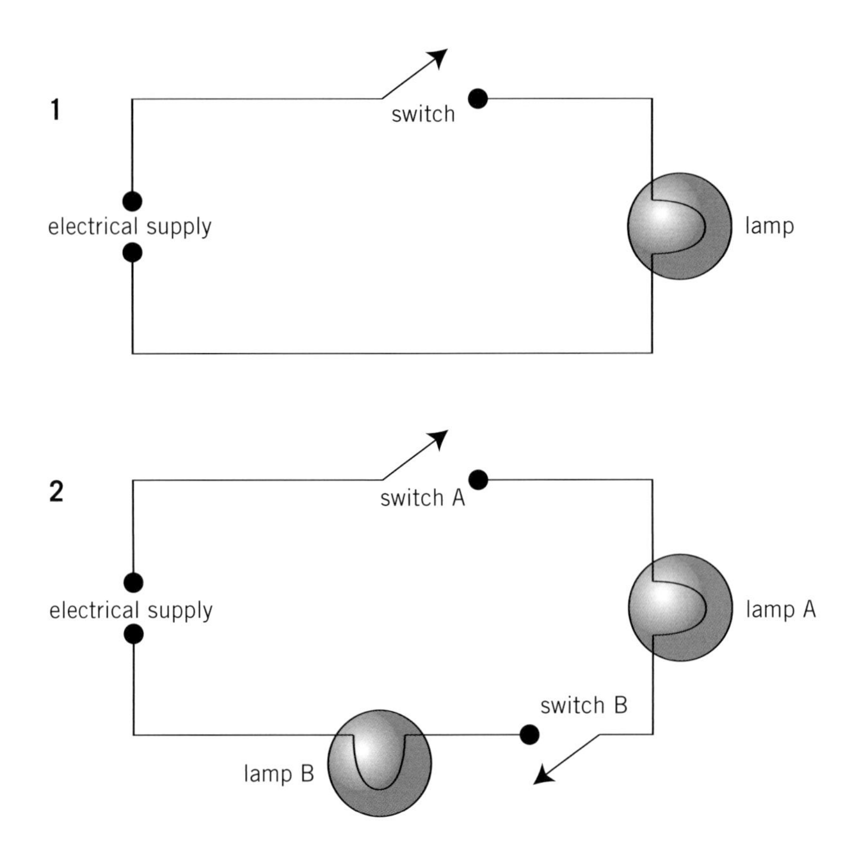

B Read the description below and match it with the correct diagram above.

The diagram above shows a basic electric circuit. It consists of an electric supply, a switch and a lamp (the electric load). The parts are connected with insulated wires (conductors). If the switch is set in the ON position, the circuit is closed and the lamp goes on. If the switch is in the OFF position, the circuit is broken and the lamp goes off.

C Complete the sentences.

1 The lamp goes on if ____

____.

2 The lamp goes off if ____

____.

D **Write a similar description for the other system.**

The diagram in exercise A shows another basic electric circuit.

____

____

____

____

E **Complete the sentences.**

1 Lamp A goes on if ____

____.

2 Lamp B goes on if ____

____.

3 Lamp B doesn't go on if ______________________________, but

____.

F **Listen and check your answers.**

G **Work with a partner. Choose a diagram from a previous page. Practise asking each other questions. Use the zero conditional.**

Lesson 5: Describing electrical systems

A **Use the information in the legend to label the diagram below.**

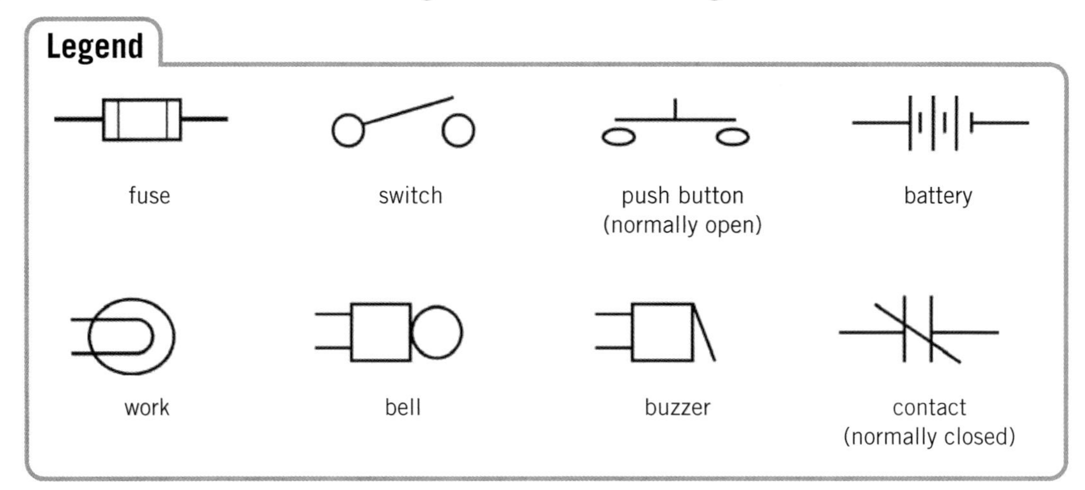

B **Discuss the diagram.**

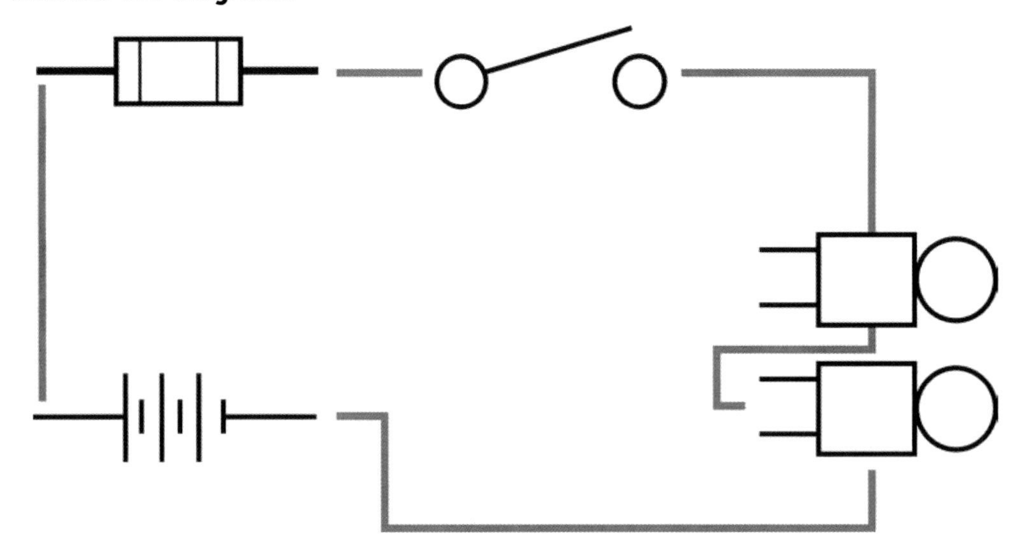

C **Match the sentence halves.**

1 This button sounds	where it makes contact.
2 How do you ring	that switch!
3 You need to charge	the buzzer.
4 The circuit is completed	a fuse.
5 Don't flick	the bell?
6 The device is operated	the battery.
7 The machine stopped working when it blew	by a push button.

D **Work with a partner.**
Student 1: Draw a basic electrical system, then describe it to your partner.
Student 2: Draw a basic electrical system, then describe it to your partner.

Draw the system your partner describes to you.

Your diagram

Your partner's diagram

Lesson 6: Using adverbs of frequency

A **The diagram below shows a network of valves in a system. Label it with the names in the box.**

gate valve	butterfly valve	three-way valve
hand-operated control valve	ball valve	normally-closed valve

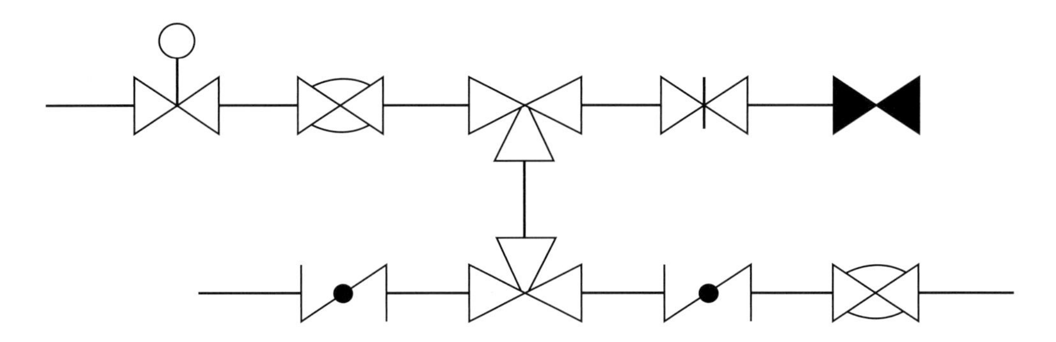

B **It is an old system and some of the valves need replacing. Look at the information below and rank the first six valves that need to be changed.**

1 The hand-operated control valve **never** leaks.

2 The first ball valve **seldom** leaks.

3 The first three-way valve **always** leaks.

4 The gate valve **usually** leaks.

5 The normally-closed valve **occasionally** leaks.

6 The second three-way valve **often** leaks.

7 The left butterfly valve **frequently** leaks.

8 The right butterfly valve **sometimes** leaks.

9 The second ball valve **rarely** leaks.

1st ______________________ 4th ______________________

2nd ______________________ 5th ______________________

3rd ______________________ 6th ______________________

C **Put the adverbs of frequency in bold from exercise B in the correct place on the scale.**

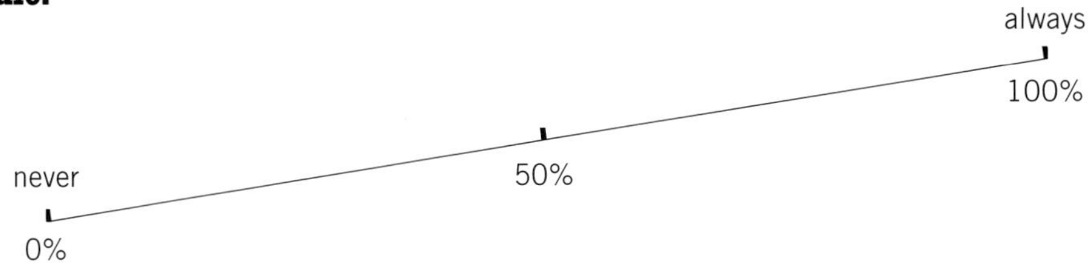

D **Interview a partner about learning English. For each question, ask for extra details, i.e., if the answer is *Never*, then you could ask why not.**

How often do you ...	Frequency	Details
study English at home?		
watch films in English?		
read things in English?		
speak in English with colleagues?		
do your homework?		
try to improve your English vocabulary?		
...		

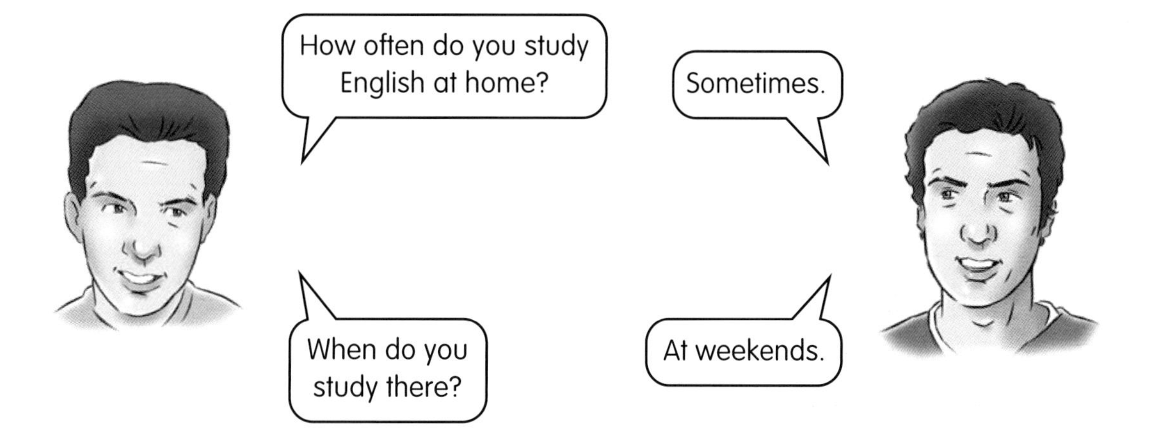

E **Work with a partner. Think of ideas of how you could improve your English outside the classroom. What are the best ways to expand your vocabulary? What opportunities do you have to use English?**

Lesson 7: Using process and instrument drawings

A **Process and instrument drawings (P & IDs) are a simple way to show complex industrial systems in diagrammatic form. Draw the correct symbol next to its definition.**

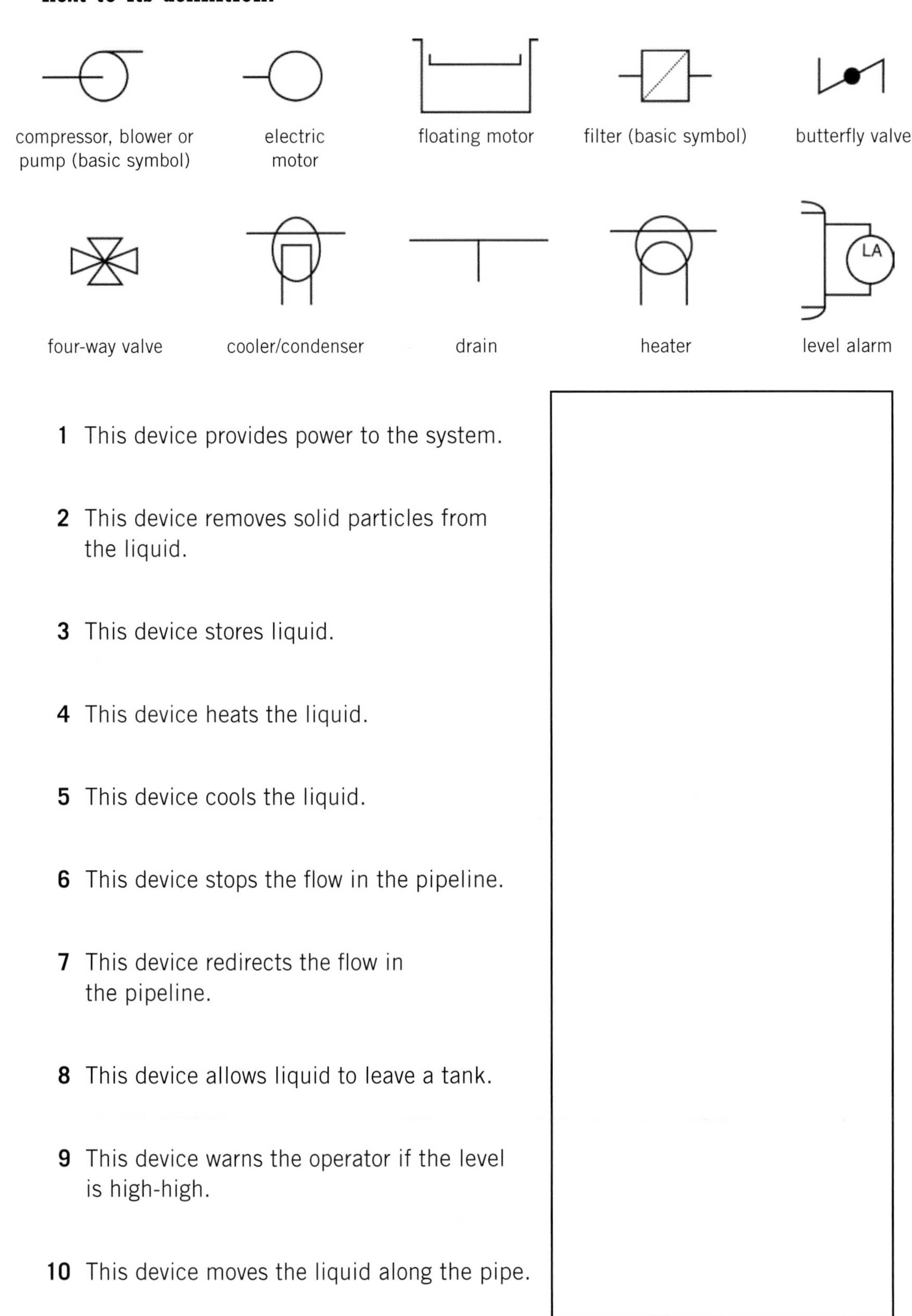

1 This device provides power to the system.

2 This device removes solid particles from the liquid.

3 This device stores liquid.

4 This device heats the liquid.

5 This device cools the liquid.

6 This device stops the flow in the pipeline.

7 This device redirects the flow in the pipeline.

8 This device allows liquid to leave a tank.

9 This device warns the operator if the level is high-high.

10 This device moves the liquid along the pipe.

B Standard abbreviations are also used to help clarify diagrams. Work with a partner.
Student 1: Look at page 220.
Student 2: Look at page 230.

Complete the table with abbreviations and full words.

1

tank	T	compressor	K
	P	filter/strainer	
heat exchanger	E		A
	C	drain	D

2

pressure	P	speed	
	F	recorder	R
level	L		I
	T	motor	

C Use the tables to work out what the abbreviations below mean.

1 LI = ____________ ____________

2 TI = ____________ ____________

3 PA = ____________ ____________

4 LA = ____________ ____________

5 TR = ____________ ____________

6 FR = ____________ ____________

7 PRI = ____________ ____________ ____________

8 SIC = ____________ ____________ ____________

9 FIC = ____________ ____________ ____________

D Listen and check your answers.

Lesson 8: Using tag numbers

Tag numbers

In addition to using symbols, P & IDs also use tag numbers. Tag numbers indicate the function of the equipment. A different tag number is used to identify each piece.

A **Look at the diagram below and practise saying the tag numbers. Which parts of the diagram can you name?**

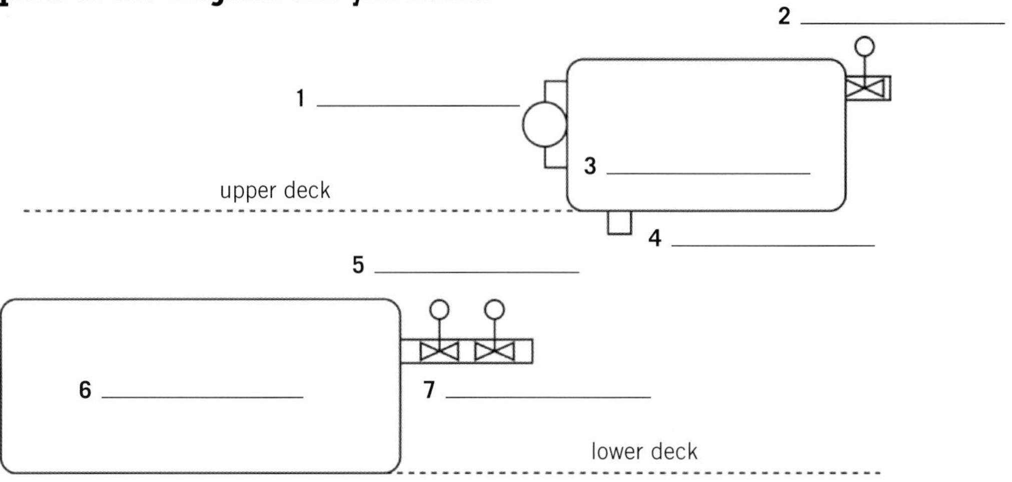

B **Listen to the description of the diagram.**

1 Label the two tanks.

2 Draw a hose to show how the tanks are connected.

C **Listen to a description of another system. Match the sentence halves.**

1 PR427-C automatically closes	for measuring high pressures.
2 The PR monitoring systems	system PR427-C2.
3 PR427-C2	if both PR427-C systems fail.
4 This leads to	valves PV576A/B.
5 The PR427-C has a back-up	initiates if PR427C malfunctions.
6 The PR427 series are suitable	are used to monitor pressure.
7 When the pressure reaches a preset	a reduction in the flow, and lowers the pressure.
8 The operator also has emergency override PRO993	PRO99 high-high level, the alarm PR427-A alerts the operator.

D **Rewrite the description in the correct order.**

The PR monitoring systems

E **Check your description with a partner.**

Lesson 9: Giving definitions

Relative pronouns

Look at how these two sentences can be changed into one sentence by using a relative pronoun.

A generator is a machine. A generator changes mechanical energy into electrical energy.

A generator is a machine ***that/which*** *changes mechanical energy into electrical energy.*

A Read the rule, then choose the correct pronoun in each sentence.

- *That* or *which* replaces the subject when it is a thing or an animal.
- *That* or *who* replaces the subject when it is a person.
- *Where* replaces the subject when it is a place.

1 An alarm is a device *which / who / that* shows that there is a problem.

2 Mostafa is an operator *which / who / that* works a four-and-two rotation.

3 A reservoir is a tank *who / that / where* water is stored.

B Rewrite each pair of sentences below as a single sentence. Use a relative clause.

1 Sarah is a lead engineer. She works on an oil rig.

____

2 A drain is a device. It allows water to leave a tank.

____

3 This is the trainer. He teaches English for the oil and gas industry.

____

4 A heater treater is a piece of equipment. Water and oil are separated there.

____

5 This is a cheap system. It is easy to install.

____

C **Complete the definitions below using a relative clause.**

1 A driller is someone who ______________________________.

2 A roustabout is a person who ______________________________.

3 An oil platform is a place ______________________________.

4 A valve is a device ______________________________.

5 Geophones are ______________________________.

D **Work with a partner. Compare your answers and decide which definitions you like best.**

E **Write definitions for the following pieces of equipment or jobs.**

1 2 3

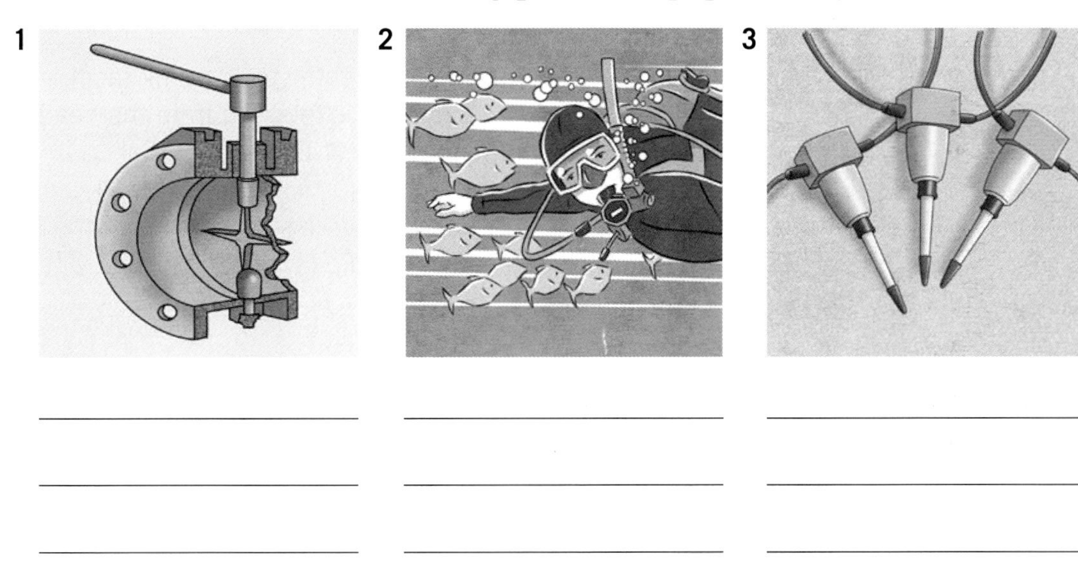

______________________ ______________________ ______________________

______________________ ______________________ ______________________

______________________ ______________________ ______________________

4 5 6

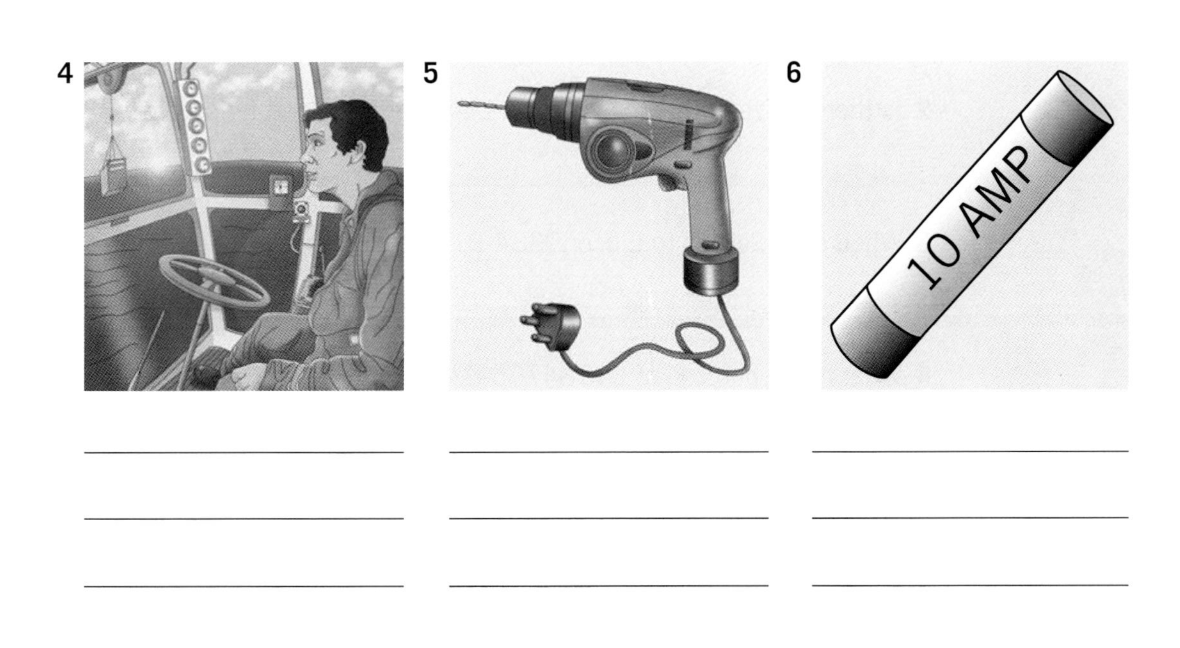

______________________ ______________________ ______________________

______________________ ______________________ ______________________

______________________ ______________________ ______________________

Lesson 10: Measuring flow

A **Listen to the definitions and guess which words in the box they are describing.**

ultrasonic	turbine	tank	back-up system	overflow	rotor

B **Listen again and complete the sentences.**

1 A __________ is a __________ which is used to hold __________.

2 An __________ is an __________ that allows excess liquid to escape.

3 A __________ is part of a mechanical or electrical __________ that __________.

C **Read the description and answer the questions.**

The Lincon9384-A and Lincon9384-B are both storage tank facilities. The Lincon9384-A is used to hold water which is released from the heater treaters. It can hold twice as much as the Lincon9384-B.

When the Lincon9384-A is full, excess water is sent to the Lincon9384-B via the pipe OP4359. OP4359 is an overflow pipe which is closed by gate valve GV384765-G during normal operations.

The Lincon9384-B also comes online as a back-up system when the Lincon9384-A is undergoing maintenance. Flow is diverted to the Lincon9384-B by three-way valve TWV85766, which can also divert flow directly to the next stage of water treatment if required.

1 What is the purpose of Lincon9384-A/B?

____

2 Is there a size difference between Lincon9384-A and Lincon9384-B?

____

3 When does flow go to Lincon9384-B?

____

4 What is the purpose of GV384765-G?

____

5 Where can TWV85766 direct flow?

____

D Read the descriptions and complete the table.

Turbine meters

At the centre of all turbine meters is a free-spinning rotor whose speed of rotation is proportional to the flow rate. As the gas or liquid passes through the pipeline, it turns the rotor, and this movement can be measured to give the flow rate. The main disadvantage of this kind of measuring device is that it has an effect on the pressure in the pipeline and disrupts the flow.

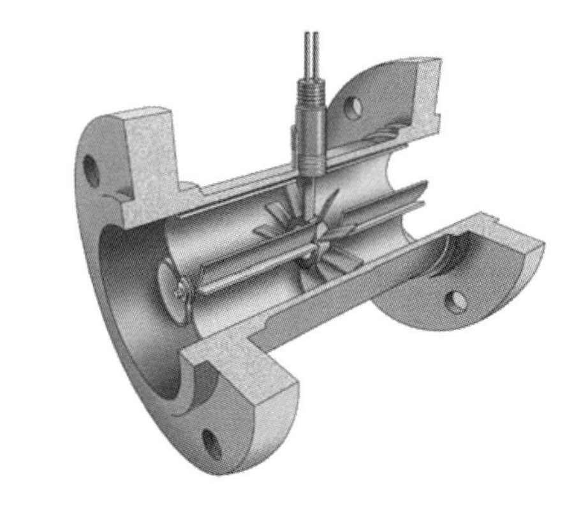

Venturi meters

These are meters that measure the pressure in the pipeline before the converging inlet, then at the throat. By calculating the difference in pressure between these two points, the flow rate can be calculated. An advantage of this kind of device is that it has little effect on downstream pressure and flow.

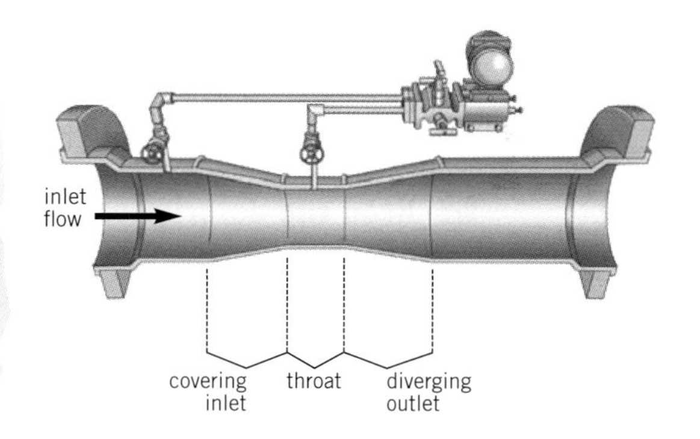

Ultrasonic meters

These meters send an ultrasonic signal through the pipeline. The flow rate can be calculated from the time it takes for a signal to be transmitted and received. This is a method which is noninvasive, and has no effect on pressure or flow rate. However, it is limited in use to fluids that can conduct ultrasound and have a well-formed flow.

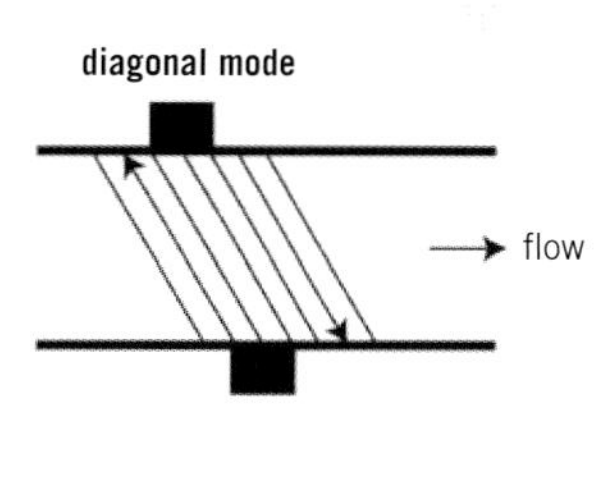

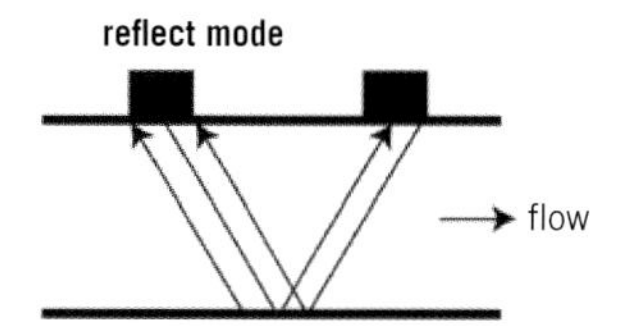

Type of meter	How flow is measured	Advantages/Disadvantages
turbine		
Venturi		
ultrasonic		

UNIT 5

Review: Describing systems

A Match each word with its definition. Then complete each definition with a relative pronoun.

1 filter	•	• An instrument ______ measures the flow of liquid in a pipe.
2 flow meter	•	• A large, open container ______ oil is stored.
3 floating roof tank	•	• Something ______ removes something from its usual position.
4 level alarm	•	• Any device ______ receives and responds to a signal.
5 displacer	•	• A device ______ removes particles from a liquid.
6 detector	•	• Something ______ gives a warning if the level of liquid is too high or too low.

B Read the description below of a pump system, then draw a diagram of the system in your notebook. Use abbreviations and P & ID symbols where necessary.

This system is powered by a motor. The motor powers a pump which forces the oil through the pipeline. The oil flows through a filter and goes into a floating roof tank. A level alarm warns the operator if the level is too high or too low. Oil can leave the floating roof tank via a drain. To prove all flow meters, we need to measure the flow under controlled conditions. This can be done by measuring the time it takes the fluid to push a displacer a known length along a pipeline. The diagram below shows a system that uses a sphere as a displacer.

C Work with a partner.
Student 1: Look at the questions on page 220.
Student 2: Look at the questions on page 230.

Ask each other questions.

UNIT 5

Assess your skills: Describing systems

Complete (✓) the tables to assess your skills.

I can ...	Difficult	Okay	Easy
• understand texts describing basic electrical systems.			
• define and explain components in a system.			
• describe how basic systems work (heating, alarm, pressure).			
• understand descriptions of systems that use abbreviations and tag numbers.			
• talk about the advantages and disadvantages of different systems.			
• describe flow measurement systems.			

I understand ...	Difficult	Okay	Easy
• the unit grammar: zero conditional			
relative pronouns			
adverbs of frequency			
defining relative clause			
• the unit vocabulary (see the glossary)			

If there is anything you are not sure of, ask your trainer to revise the material.

UNIT 6 TALKING ABOUT SAFETY

The aim of this unit is to:

- provide the language and structures to talk about safety, both in terms of identifying hazards and reporting incidents

By the end of this unit, you will be able to:

- identify parts of the body
- identify different types of injury
- identify and explain the use of personal protective equipment (PPE)
- identify safety signs
- identify hazards
- report incidents using appropriate language

Lesson 1: Talking about parts of the body and injuries

A Complete the table with the words in the box.

toe neck thumb finger back
shoulder chest elbow ankle
ear face leg nose mouth
eye hand foot knee arm wrist

Torso	Head

Upper limbs	Lower limbs

B Label the diagram with the correct words.

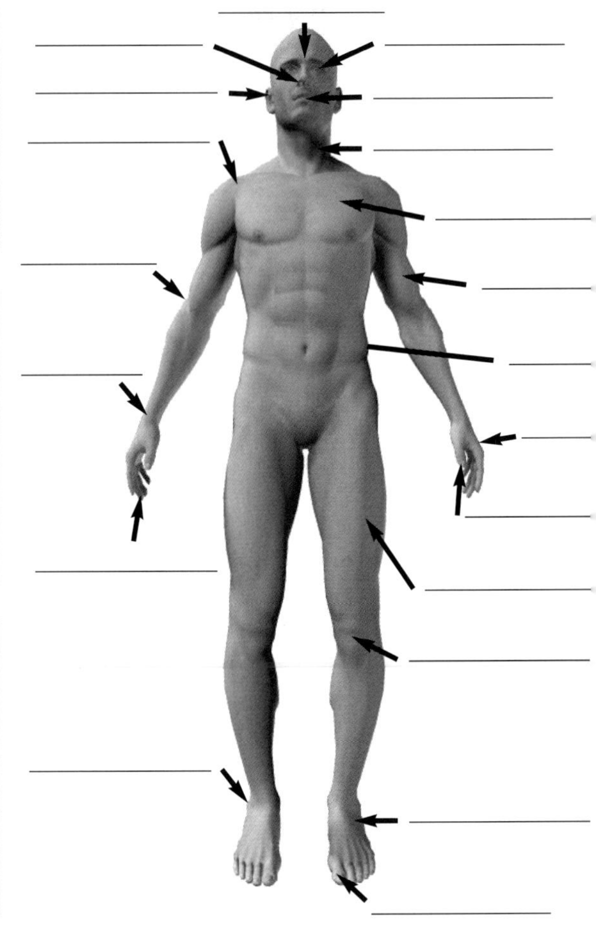

C **Read the dictionary definitions of different types of injury. Discuss which injuries are shown in the pictures. Note that more than one answer is possible for one picture.**

contusion (*n*): an injury where the skin is not broken, such as a bruise,
e.g., *The victim had contusions around the neck.*

dislocation (*n*): when a bone or joint is moved out of place,
e.g., *A dislocated shoulder is very painful.*

fracture (*n*) (*v*): a break in a bone,
e.g., *She fractured her collar bone when she fell.*

laceration (*n*): a wound where the skin is cut or torn,
e.g., *He lacerated his finger with the knife.*

sprain (*n*) (*v*): a twist of a joint, or tearing of the soft tissue around a joint,
e.g., *His ankle was sprained, but not broken.*

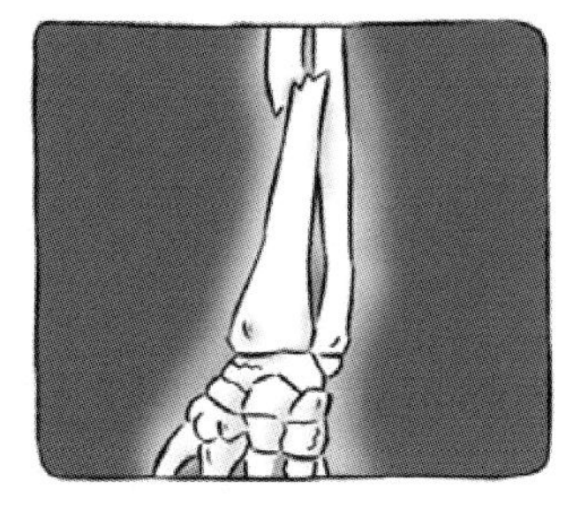

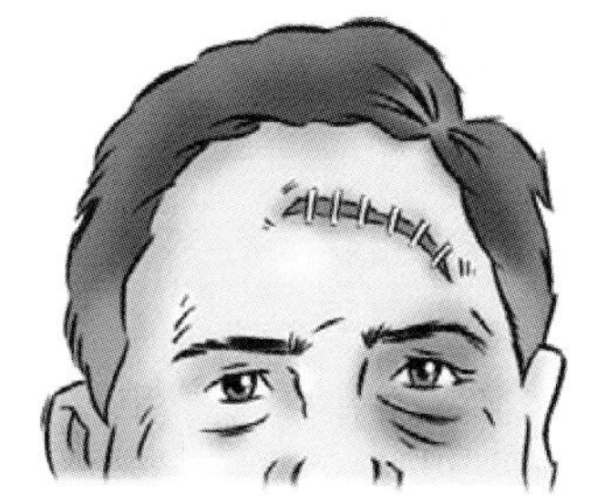

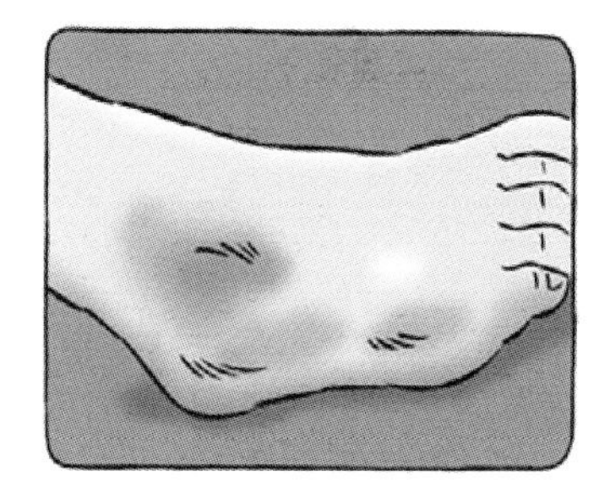

1 ______________ 2 ______________ 3 ______________

D **Alan was in an accident. Listen and circle the body parts he hurt.**

E **Listen again and complete the accident report form.**

Date and time of accident	
Place of accident	
Reason for accident	
Injuries	

Lesson 2: Talking about PPE and safety equipment

A Read and complete the signs.

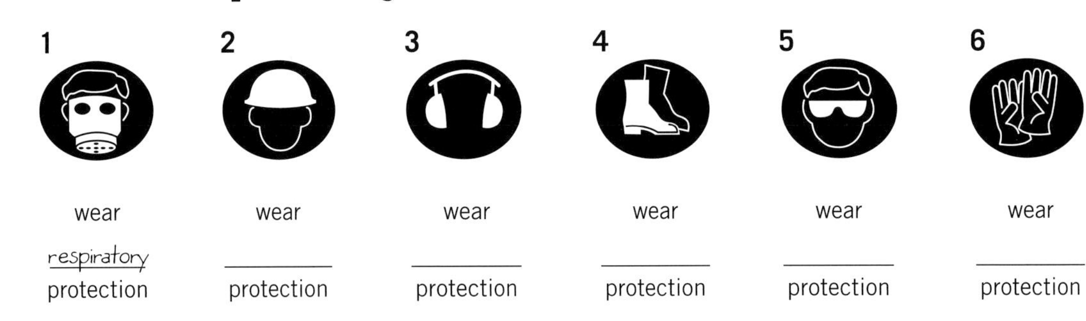

B Identify the PPE in the pictures below and say what type of protection it is. Use the words in the box.

safety	boots	hat	shoes	mask	gloves	hard	goggles
face	protectors	overalls	ear				

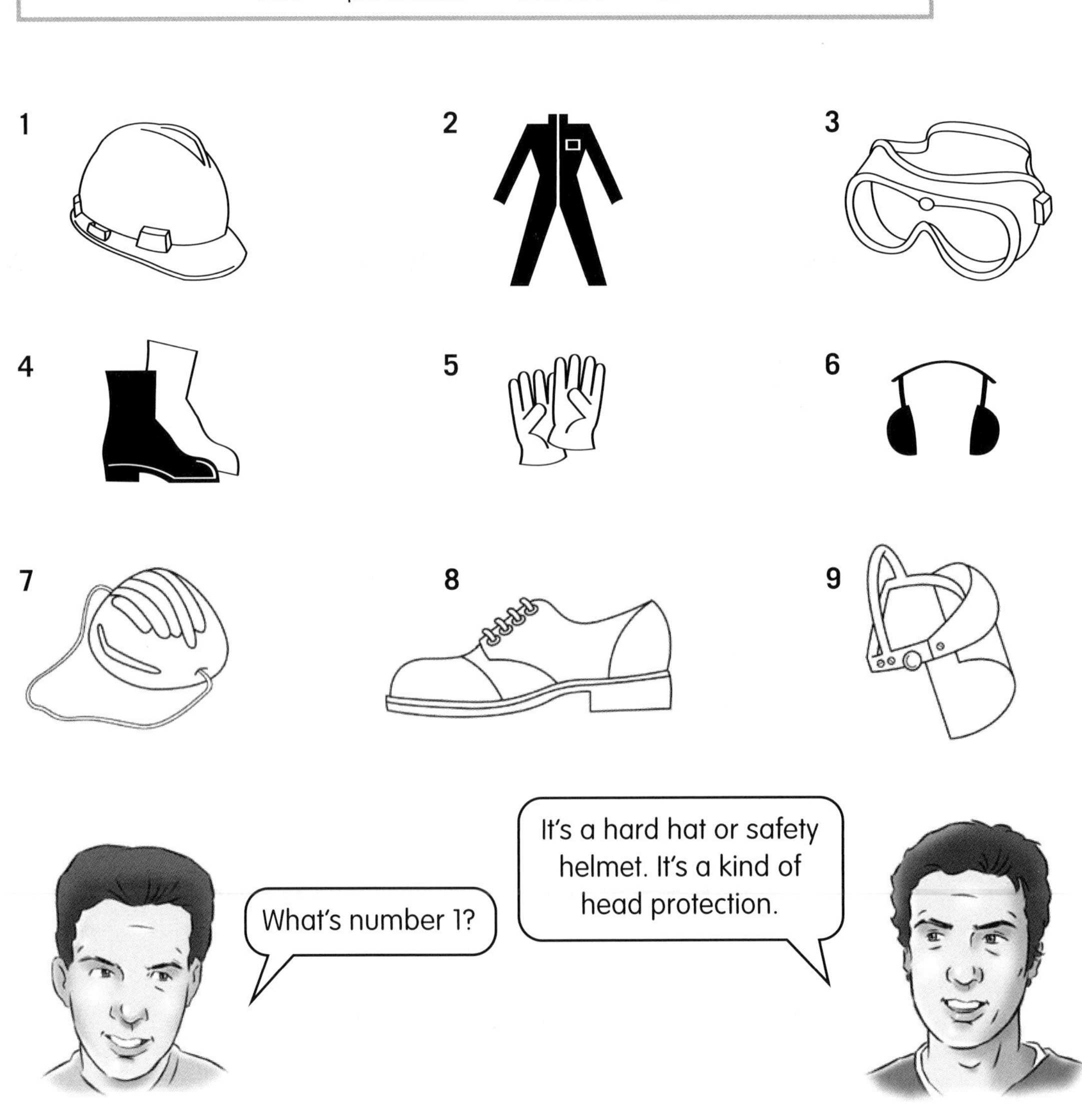

C Match each item of safety equipment with its definition.

1 machine guard •	• An inflatable device that floats, which you throw to someone in the water.
2 first-aid kit •	• A signal that produces a bright flame to attract attention.
3 fire extinguisher •	• A type of gate or fence that stops people going somewhere.
4 safety barrier •	• A rope used for saving people in danger, especially at sea.
5 flare •	• A box containing medicines and bandages to treat injured people.
6 life buoy •	• A set of bands worn to hold someone, or stop them from falling.
7 lifeline •	• An attachment or a covering on a machine to protect the operator or a part of the machine.
8 safety harness •	• Orange and white objects that are placed around an area to stop people going there.
9 safety cone •	• A metal container containing water or chemicals for stopping fires.

D Name the equipment in the pictures. Discuss where you can see each item.

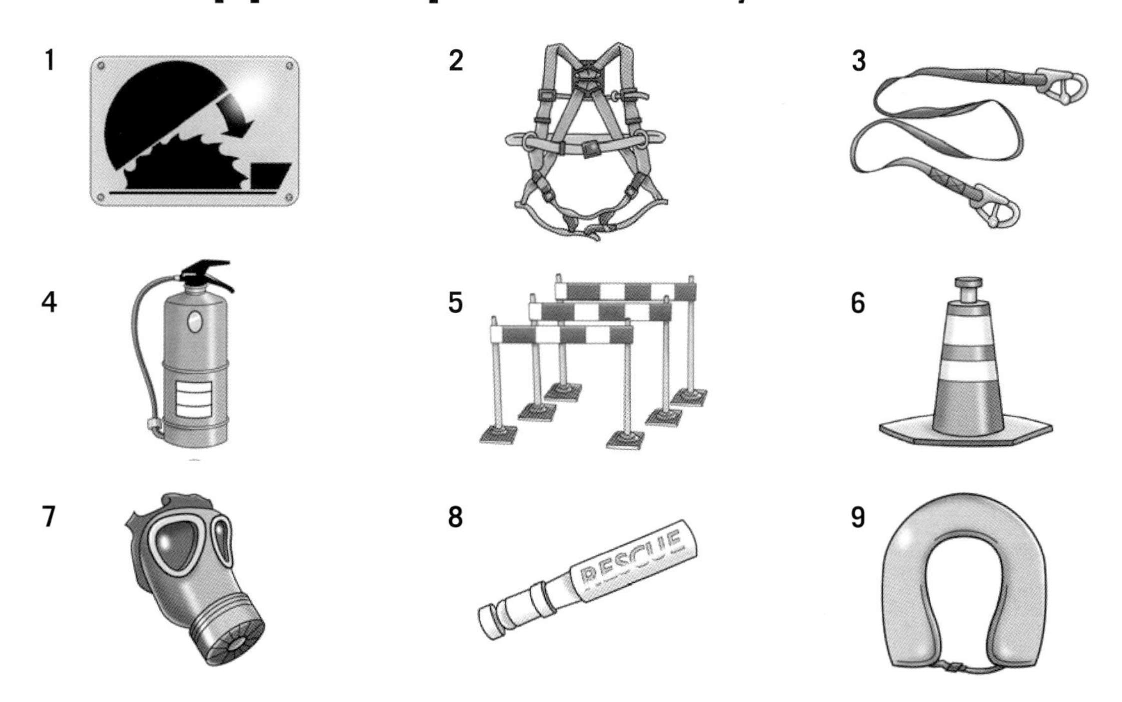

E Listen and identify the equipment above that people are discussing.

Lesson 3: Using different parts of speech

A Read the text and complete the table.

Hands and fingers are the part of the body most at risk of injury. Your hands are your principal tools, and it is important to look after them. One way to protect your hands is to wear appropriate gloves. There are different types of glove available that are suitable for different jobs. Disposable gloves protect the hands against mild irritants, but chemical-resistant gloves protect hands against hazardous chemicals. Fabric gloves are lightweight gloves that improve grip so you can hold slippery objects. They are also useful for protecting hands against mild heat or cold. Leather gloves are heavy-duty gloves that offer more protection. They insulate the hands against heat and cold, and they are also able to protect the hands when working with sharp objects or rough edges. However, no pair of gloves can protect you from carelessness.

Type of glove	Heat/Cold	Mild irritants	Hazardous chemicals	Carelessness	Sharp edges
disposable		✓		✗	

B Look at the information above, then decide which types of gloves are suitable for the following situations.

1 Working outside in winter __________

2 Handling cut pieces of pipe __________

3 Handling barrels of chemicals __________

4 Cleaning an oil spill __________

5 Using a cutting machine __________

6 Handling hazardous chemicals __________

7 Handling greased machine parts __________

C Identify the form of the words in the boxes. Complete the sentences with the correct form of the word.

safe *adjective*	unsafe ________	safety ________

1 An ________ action is one that can cause an accident or injury.

2 ________ is everybody's responsibility.

3 Always check conditions are ________ before you start work.

hazard ________	hazardous ________

1 A ________ is a possible danger.

2 Sulphuric acid is a ________ chemical.

protect ________	protector ________
protection ________	protective ________

1 It is important to wear the correct personal ________ equipment (PPE) in the workshop.

2 Fabric gloves ________ your hands against mild heat or cold.

3 Leather gloves offer stronger ________.

4 Helicopter pilots must wear ear ________.

dispose ________	disposable ________	disposal ________

1 What do you do about waste ________?

2 Please ________ of your gloves in the yellow bin.

3 ________ gloves do not protect against heat.

careless ________	careful ________	carelessness ________

1 If you see a ________ act, you should talk to the person.

2 Always be ________ at work.

3 ________ causes accidents.

Lesson 4: Talking about hazards and causes of accidents

A **What do the safety signs mean? Use some of the words in the box.**

hazard	trip	emergency exit	ear protectors	danger	first-aid	
emergency	exit	no admittance	PPE	high voltage	permit	stop
safety	gloves	electric shock	forklift truck	mobile phones	required	

1

2
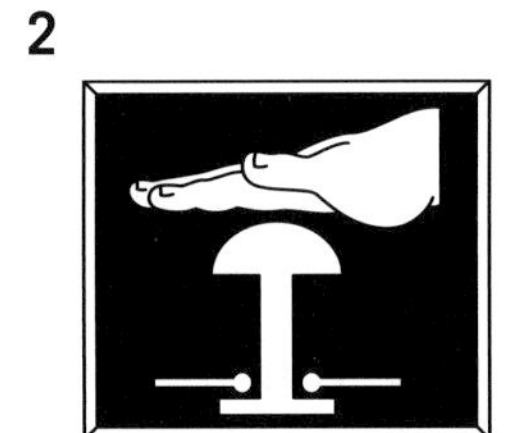

3

4

5

6
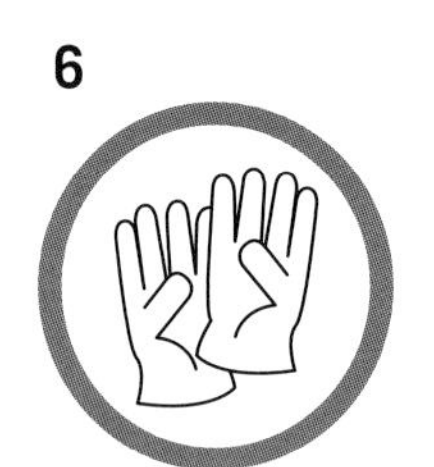

7
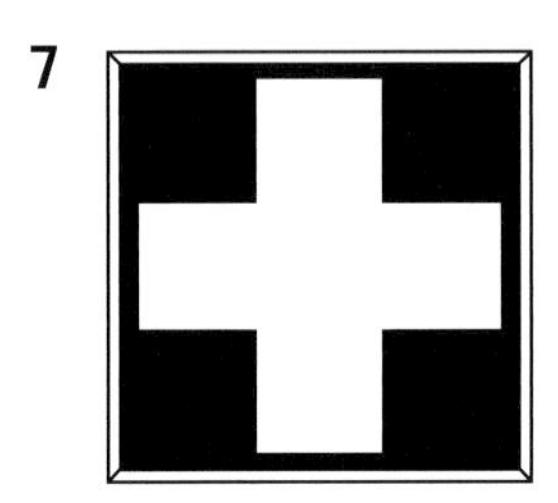

8

9

10
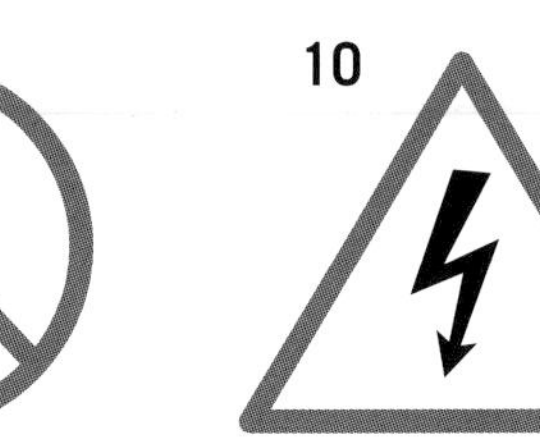

11

12

B Match the sentence halves.

1 If there are cables on the floor,	find the nearest emergency exit.
2 If there is a fire,	make sure you have a permit.
3 If you travel by plane,	be careful not to trip.
4 If you go into a restricted area,	make sure your mobile phone is turned off.

C Make full sentences using an *if* clause and an imperative.

1 If / you / not have / permit / not / come / in / !

____

2 If / someone / has / electric shock / get / help

____

3 Press / emergency stop button / if / problem / machine

____

4 Watch out / forklift trucks / if / go / Area D

____

D Write more sentences. What should you do if ...

1 a cable lies across an aisle?

____

2 someone has an accident?

____

3 someone spills water on the ground?

____

4 a ladder is at the wrong angle?

____

5 a person does not use a guard on a grinding machine?

____

6 it is very windy?

____

Lesson 5: Talking about risk assessment

A **Look at the picture and identify possible hazards.**

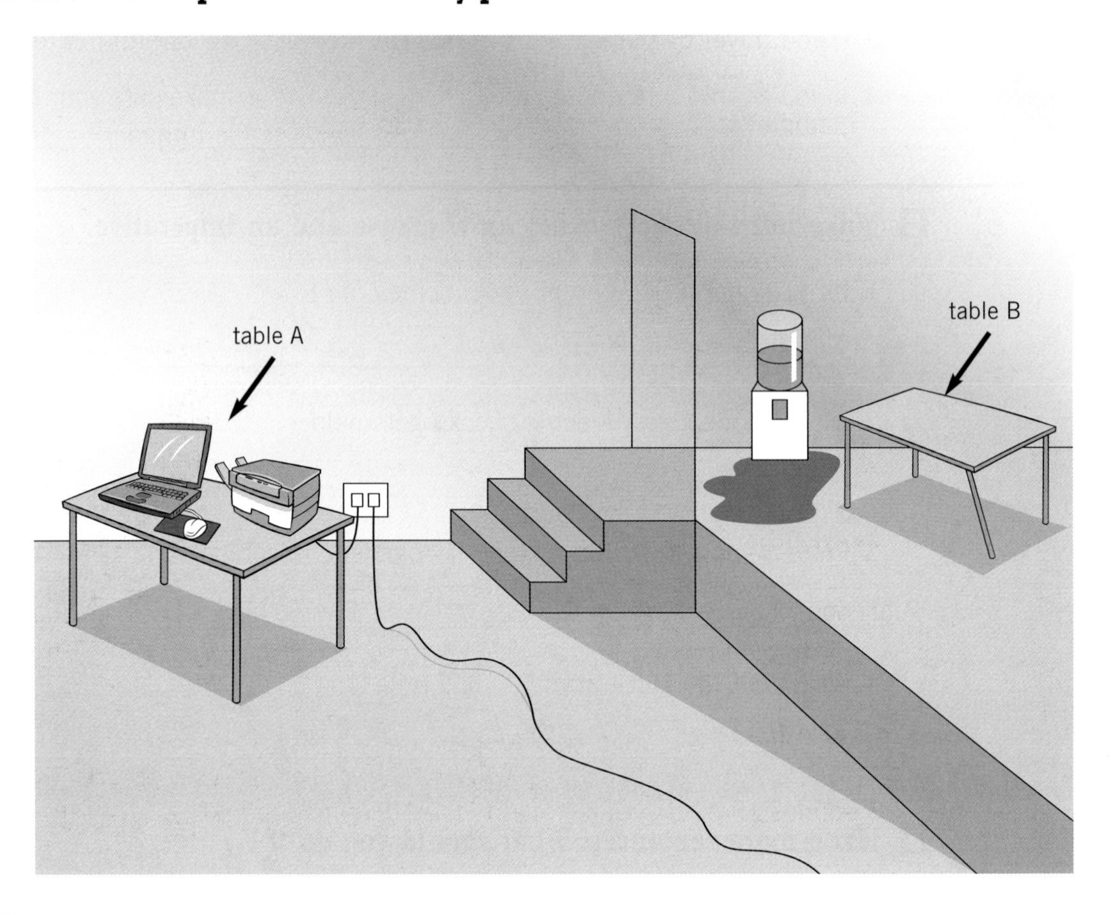

B **Bob wants to move the printer from table A to table B. Listen to his risk assessment and complete the sentences.**

1 If we ________________ the printer on table B, it ________________ probably collapse.

2 If we ________________ the cable, someone ________________ over it.

3 Someone ________________ if we ________________ up the spilt water.

4 If you ________________ a heavy object, ________________ your legs before you lift it.

C Match the forms with the example sentences in exercise B.

First conditional

These sentences talk about specific possible situations in the future and are formed by either:

If + present tense, + *will* + base form.

Example: ______________________________

If + present tense, + other modal auxiliary + base form.

Example: ______________________________

If + present tense, + imperative verb form.

Example: ______________________________

We make the negative by adding *don't*.

Example: ______________________________

D Rearrange the words to make sentences.

1 won't / the / rains / it / helicopter / if / fly / .

The helicopter won't fly if it rains. /

If it rains, the helicopter won't fly.

2 sounds / fire / evacuate / alarm / the / building / the / if / .

____

____

3 follow / not / could / have / do / procedures / you / you / accident / an / safety / if / .

____

____

4 weather / complete / the / the / if / good / will / we / is / job / .

____

____

E Discuss what you will do if ...

1 someone can't understand your English.

2 you have some free time next week.

3 you have a lot of homework.

4 you see someone doing something dangerous.

Lesson 6: Talking about past events

A Listen to Ahmed and Bob talking. What was the problem?

B **Work with a partner.**
Student 1: Complete the past simple verb table on page 220.
Student 2: Complete the past simple verb table on page 230.

Check your answers with your partner.

1 Which verbs add *-ed* to form the past simple tense?

2 Which verbs add *-d*?

3 Which verbs double the final consonant and add *-ed*?

4 Which verbs have an irregular past simple tense form?

5 Do you notice any other patterns?

C **Complete the table with verbs from exercise B.**

Verbs that add *-ed* to form the past simple	
Verbs that add *-d* to form the past simple	
Verbs that double the final consonant and add *-ed*	
Irregular verbs	

D **Read the summary and underline the verbs with spelling mistakes. Then rewrite them correctly.**

Bob comed to work at 7.00 a.m. He heared the alarm so he checked the control panel. The flow rate were too low. He tryed to increase the flow rate by opening all the valves, but it didn't solve the problem. He calld the pump station and they increaseed the pressure.

____________________ ____________________

____________________ ____________________

____________________ ____________________

E **Look at how the negative and the question forms of the past simple are formed. Then choose the correct option to complete the sentences below.**

*She finish**ed** the job on time.*
*She **didn't** finish the job on time.*
***Did** she finish the job on time?*

1 I *not heard / didn't heard / didn't hear* the alarm.

2 What *did / do / have* you *do / did / done* yesterday?

F **Complete the conversation between Hassan and his supervisor. Put the verbs in brackets into the past simple.**

Tom: What happened yesterday?

Hassan: I [1] __________ (*turn on*) the monitoring equipment and [2] __________ (*check*) the system as usual. Then, at 11 o'clock, an alarm [3] __________ (*sound*).

Tom: What was the matter?

Hassan: The flow rate [4] __________ (*be*) too high.

Tom: Did you initiate an emergency shutdown?

Hassan: No, I [5] __________ (*not/initiate*) an emergency shutdown. I [6] __________ (*adjust*) the flow rate and [7] ________ (*lower*) the pressure in the pipeline. The alarm [8] ________ (*stop*) when the flow rate [9] ________ (*reach*) normal levels.

G **Listen and check your answers.**

Lesson 7: Talking about actions in progress in the past

A Choose the correct sentence for each picture.

1 At 8.00 p.m. she was driving home. ☐

2 At 8.00 p.m. she drove home. ☐

1 They left the workshop when the phone rang. ☐

2 They were leaving the workshop when the phone rang. ☐

B **Read about the past continuous and complete the sentences.**

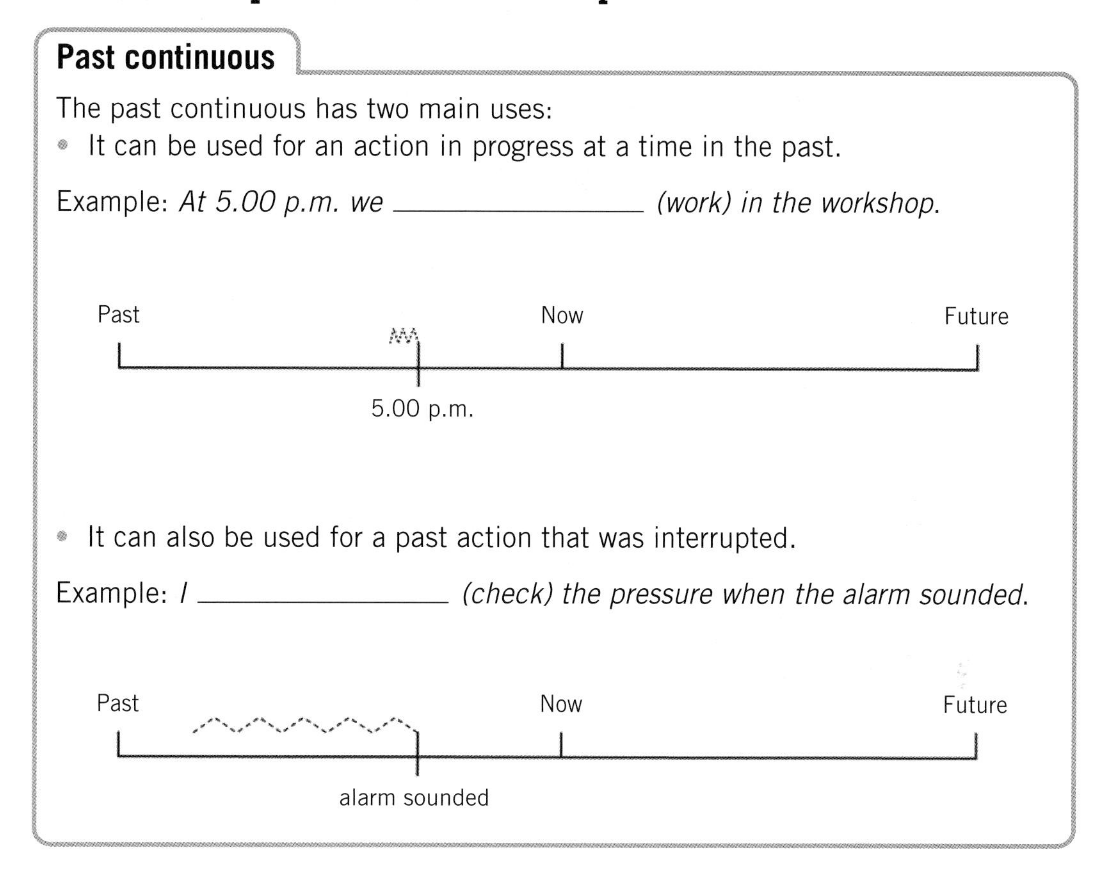

Past continuous

The past continuous has two main uses:

- It can be used for an action in progress at a time in the past.

Example: *At 5.00 p.m. we* ______________ *(work) in the workshop.*

- It can also be used for a past action that was interrupted.

Example: *I* ______________ *(check) the pressure when the alarm sounded.*

C **Answer the questions using the past continuous.**

1 What were you doing at this time yesterday?

____

2 What were you doing before you started this course?

____

3 What was happening when you arrived here today?

____

4 What was happening in your country in 2005?

____

5 Have you ever had an accident?

____

6 What were you doing when it happened?

____

Lesson 8: Reporting incidents

A Choose the best definition for the underlined words.

1 We will observe the group.
- **a** to watch carefully
- **b** to obey

2 I encouraged him to talk about it.
- **a** to prevent
- **b** to persuade

3 He is responsible for safety.
- **a** in charge of
- **b** worried about

4 We need to modify the truck.
- **a** to make stronger
- **b** to make small changes

5 We need to eliminate the problem.
- **a** to reduce
- **b** to get rid of

6 We have clear procedures.
- **a** rules for how to do things
- **b** movements

B Complete the text with the words above.

Safety Training Observation Programme

The STOP system is based on the following idea. Everyone is [1]____________ for safety, and all injuries can be eliminated if safe [2]____________ are reinforced. STOP report cards can be positive or negative because people who work safely should be told, as well as those who do not. The key to reducing incidents and injuries is to [3]____________ people, talking with them to [4]____________ safe work practices and, therefore, [5]____________ their behaviour to [6]____________ unsafe acts and behaviours.

C Answer the questions.

1 Who is responsible for safety?

____

2 Why should you talk to someone who does an unsafe act?

____

3 What can happen if safe procedures are reinforced?

____

D Complete the STOP report cards with the verbs in the correct tense. Then listen and check.

1 A technician [1] __________ (*notice*) a piece of wood with nails sticking out in front of the equipment lockup. First, he [2] __________ (*bend*) the nails over with a hammer, then [3] __________ (*put*) the wood in the garbage container. I [4] __________ (*congratulate*) him on his action.

2 While I [1] __________ (*wash*) my hands, I [2] __________ (*notice*) a broken mirror frame in the bathroom, which [3] __________ (*be*) in danger of falling. I [4] __________ (*inform*) Administration, and the company [5] __________ (*repair*) the frame.

3 The light in a classroom [1] __________ (*malfunction*) before lunch. The light [2] __________ (*flash*) on and off. I [3] __________ (*think*) it [4] __________ (*be*) a short circuit and a fire hazard. I [5] __________ (*notify*) Administration and they immediately [6] __________ (*send*) someone to repair the light.

E Write a STOP card report about one of the pictures below. Remember to say where the incident happened, what you saw, what you said to the person, and what happened after.

1

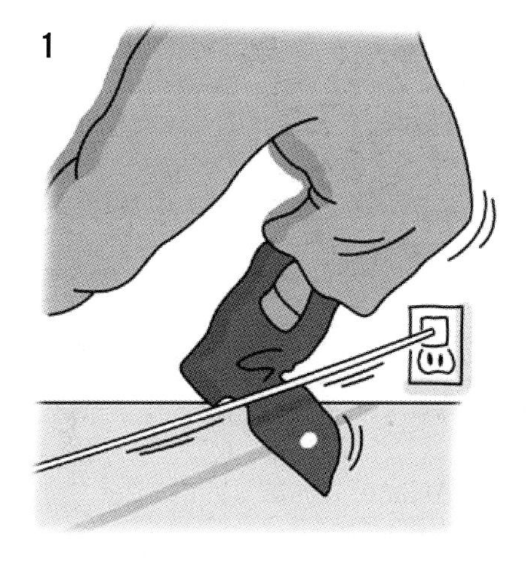

2

______________________________ ______________________________

______________________________ ______________________________

______________________________ ______________________________

Lesson 9: Asking about incidents

A Complete the table with appropriate question words.

Question word	Use in questions about ...
What ...?	things
	place
	time
	people
	reason
Which ...?	choice
	ownership
	method
How often ...?	frequency
	quantity
	age
	length of time
	distance

B Complete the dialogue.

A: What's this? **B:** It's a valve.

A: ____________________? **B:** On the left side of the tank.

A: ____________ was it installed? **B:** It was installed two years ago.

A: Who __________________? **B:** By a maintenance technician.

A: ____________________? **B:** Because the old one was faulty.

A: How __________________? **B:** He isolated the section of pipe, removed the cover and replaced the valve.

A: __________ did it take to replace? **B:** Just under thirty minutes.

C Listen to Bob describing an incident. Think of questions you would like to ask about it.

____

____

____

____

____

____

____

____

____

D Listen to Ahmed asking Bob about the incident. Were his questions the same as yours?

E Work with a partner. Ask each other questions about the incident. One person plays the role of Bob and one person plays the role of Ahmed.

Lesson 10: Talking about the golden rules

A **Discuss in groups what you think are important safety rules for people working in the oil, gas and petrochemicals industries.**

B **Read the introduction to BP's 8 golden rules. Find words in the text that mean the same as the words and phrases below.**

Getting the basics right

BP's safety policy states no harm to people and no accidents. Everyone who works for, or on behalf of, BP is responsible for their safety and the safety of those around them.

The following safety rules will be strictly enforced to ensure the safety of our people and our communities.

Although embedded in each of these rules, it is important to emphasize that:

- work will not be conducted without a pre-job risk assessment and a safety discussion appropriate for the level of risk.
- all persons will be trained and competent in the work they conduct.
- personal protection equipment will be worn as per risk assessment and minimum site requirements.
- emergency response plans, developed from a review of potential emergency scenarios, will be in place before commencement of work.
- everyone has an obligation to stop work that is unsafe.

1 injury or damage (n): ______________________

2 made to happen (v, past participle): ______________________

3 analysis of possible dangers: ______________________

4 having the skill to do something well: ______________________

5 things that must be done on the site: ______________________

6 plans of how to react in a dangerous situation: ______________________

7 start (v): ______________________

8 duty ______________________

C **Write each safety rule in the box next to its explanation.**

Energy isolation	Management of change (MOC)	Ground disturbance
Working at heights	Driving safety	Lifting operations
Permit to work	Confined space entry	

BP's 8 golden rules of safety

1 ____________________: Before conducting work that involves confined space entry, work on energy systems, ground disturbance or hot work in potentially explosive environments, a permit must be obtained. This should be authorized by a responsible person; it should also identify hazards, assess risk and establish control measures.

2 ____________________: Working at heights of two metres or higher above the ground cannot begin unless the workers are competent to do the work, and the correct equipment is in place and has been inspected.

3 ____________________: An isolation of energy systems, mechanical, electrical, process, hydraulic and others, cannot proceed unless the method and discharge of stored energy are agreed and executed by a competent person. They should also test that the isolation is effective and continue to monitor effectiveness periodically.

4 ____________________: Entry into any confined space cannot proceed unless all other options have been ruled out and the area has been tested for safety. All affected personnel, including a stand-by person, should be competent and issued with a permit.

5 ____________________: Lifts utilizing cranes, hoists, or other mechanical lifting devices will not commence unless an assessment of the lift has been completed. The equipment must be certified for use and examined before each lift. The operators must be trained, and the rigging of the load must be done by a competent person.

6 ____________________: Work arising from temporary and permanent changes to organization, personnel, systems, process, procedures, equipment, products, materials or substances, and laws and regulations cannot proceed unless a Management of change process is completed. This should include a risk assessment conducted by all affected by the changes, authorization of the changes and a clear work plan for the changes.

7 ____________________: All categories of vehicle, including self-propelled mobile plant, must not be operated unless the vehicle has been inspected and is in good working order, and safety equipment (such as helmets and seat belts) is used. Drivers must be trained and certified, should not use radios or hand-held phones and should not be under the influence of drugs and alcohol.

8 ____________________: Work that involves a manmade cut, cavity, trench or depression in the Earth's surface formed by earth removal cannot proceed unless a hazard assessment is completed and all underground hazards have been identified and isolated where necessary.

(adapted from BP's Golden Rules of Safety and used with the permission of BP plc)

D **How similar or different are these rules to the rules you thought of? Discuss which ones you think are most relevant or important to you.**

UNIT 6

Review: Talking about safety

A Identify the potential hazards in the picture and discuss them with a partner.

B Work with a partner.
Student 1: Read the incident report on page 221.
Student 2: Read the incident report on page 231.

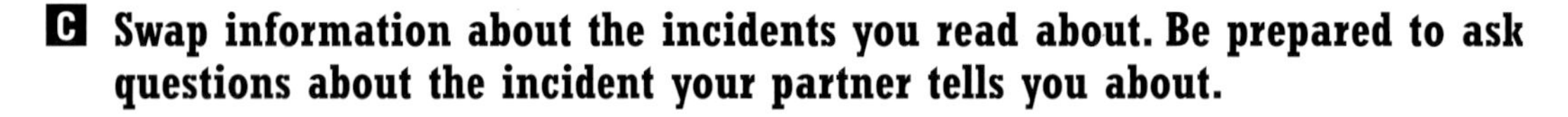

C Swap information about the incidents you read about. Be prepared to ask questions about the incident your partner tells you about.

D Complete the incident report form.

Date and time of incident	
Where the incident took place	
Description of incident	
Action taken	

UNIT 6

Assess your skills: Talking about safety

Complete (✓) the tables to assess your skills.

I can …	Difficult	Okay	Easy
• name parts of the body.			
• name types of injury.			
• name and explain the use of safety equipment.			
• identify and talk about potential hazards.			
• describe incidents in the past.			
• ask and answer questions about past incidents.			
• complete incident reports.			
• understand and discuss safety rules.			

I understand …	Difficult	Okay	Easy
• the unit grammar: parts of speech			
first conditional			
past simple			
past continuous			
question words			
• the unit vocabulary (see the glossary)			

If there is anything you are not sure of, ask your trainer to revise the material.

UNIT 7

MAKING COMPARISONS

The aim of this unit is to:

- practise the functions of comparison and evaluation

By the end of this unit, you will be able to:

- use comparative and superlative structures
- express similarities between things
- use quantifiers to show the degree of difference between things
- discuss the qualities of metals
- identify the materials that objects are made from
- discuss the oil-refining process

Lesson 1: Making general comparisons between two things

Comparative adjectives

When we want to compare two or more objects, we use comparatives. Comparatives are made with adjectives.

A　　B

A is bigger than B.　　*B is smaller than A.*

A Complete the tables with the adjectives in the box and their comparatives.

big useful heavy flexible fast expensive long bad easy dirty hot safe good dangerous complex

Short adjectives				
One-syllable	**Ending in C V C**	**Ending in *-e***	**Ending in *-y***	**Irregular**
cold – colder	fat – fatter	close – closer	pretty – prettier	far – further/farther

Long adjectives
Two- or three-syllable
practical – more practical

B **Complete the sentences with the adjectives from exercise A in the comparative form.**

1 An electric drill is ______________________ a hand drill.

2 A mobile phone is ______________________ a landline phone.

3 Rubber is ______________________ steel.

4 A job offshore is ______________________ a job in an office.

5 Lead is ______________________ wood.

6 A mile is ______________________ a kilometre.

7 A jet engine is ______________________ a combustible engine.

C **Work in small groups. Look at the pictures of two types of recovery. Discuss what you know about each method.**

nodding donkey

offshore platform

D **Listen and circle the words and phrases that are used to compare the different types of oil recovery.**

sophisticated dangerous small expensive reliable cheap to run
easy to maintain effective traditional simple useful drills deeply
modern practical

E **Write six sentences comparing methods of oil recovery. Use comparative forms.**

Lesson 2: Making more specific comparisons

A **Look at the diagram and identify what it is. Finish labelling the diagram with the words in the box.**

~~hole wall~~ drill bit casing drill string surface

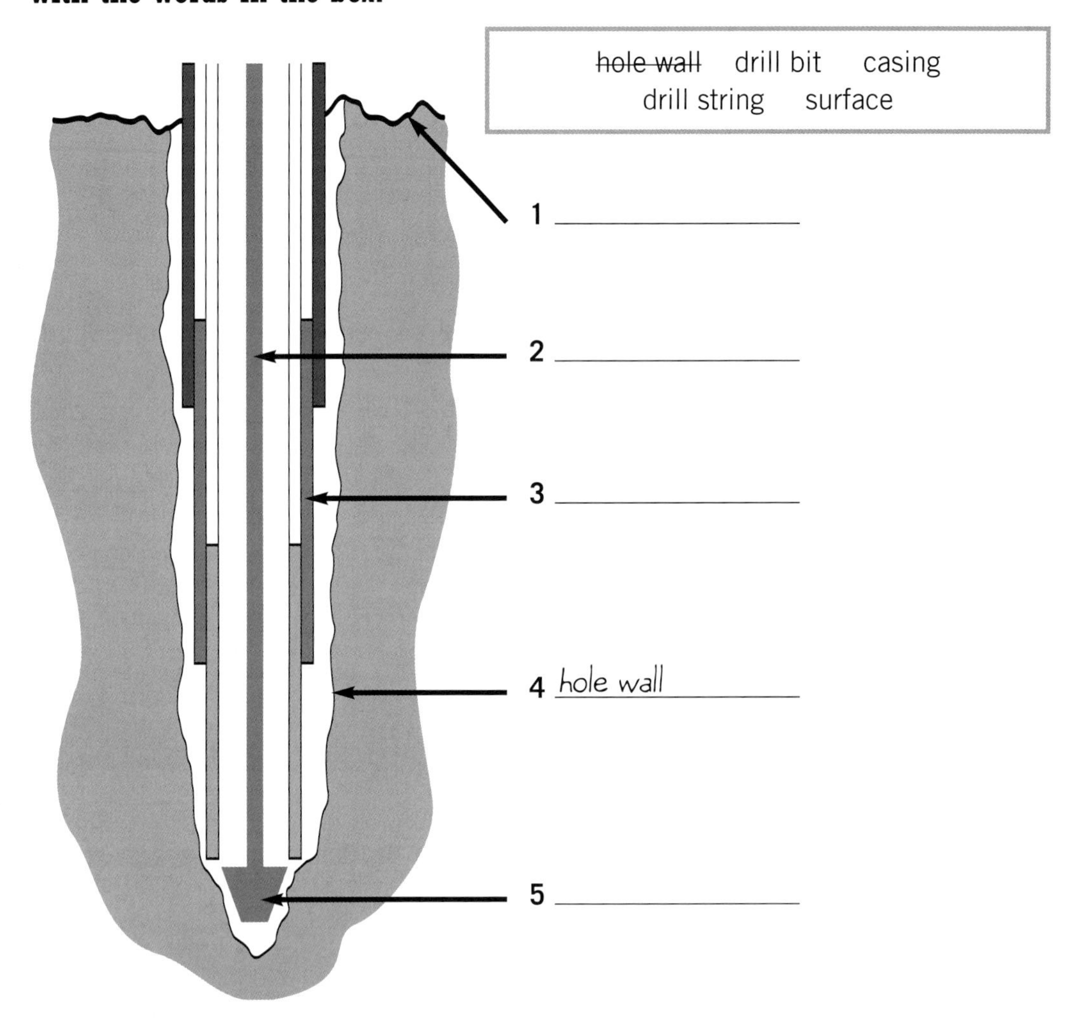

B **Listen and check your answers.**

C **Read the sentences and underline the mistakes in the use of comparatives. Then rewrite them correctly.**

1. Because deep formations have more higher pressures than shallow formations, well casings are needed to protect weakker upper formations.
2. When a well is drilled, the top diameters are wideer than those deeper in the well.
3. The casing diameter is more narrower at the bottom of the well than at the top.
4. The hole is always slightly biger than the casing so that a cement bond can be pumped between the outside of the casing and the wall of the hole.
5. After the first section of well is drilled, a wide diameter casing is fitted inside the hole. A drill bit more thin than the casing is then used to drill the next section of the hole.

Adverbs of degree

We can make comparatives more specific if we add an adverb of degree to clarify whether there is a big or small difference between the objects that are being compared.

D Complete the sentences.

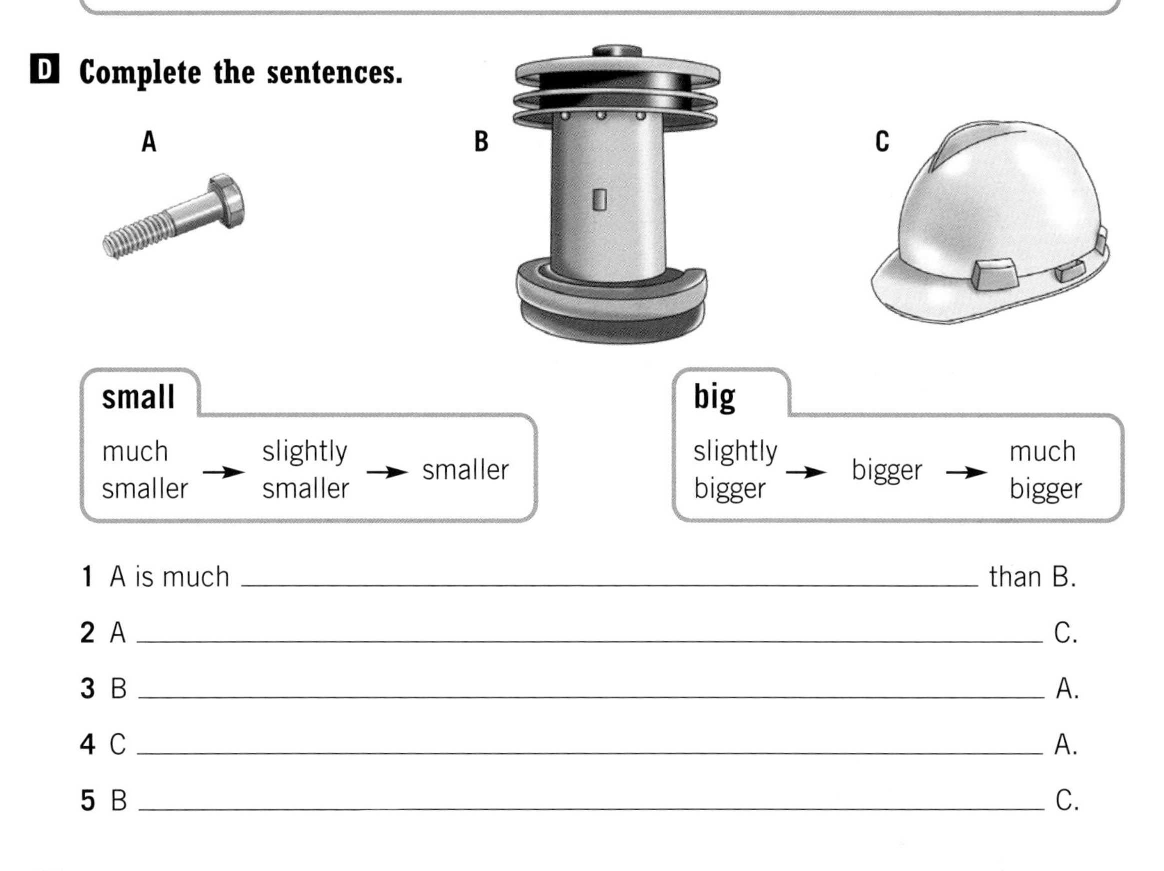

small

much smaller → slightly smaller → smaller

big

slightly bigger → bigger → much bigger

1 A is much ______________________________ than B.

2 A ______________________________ C.

3 B ______________________________ A.

4 C ______________________________ A.

5 B ______________________________ C.

E Listen and write down two adjectives that are used to describe each liquid, e.g., *viscous*, *heavy*, etc.

F Work with a partner. Take it in turns to compare the different liquids.

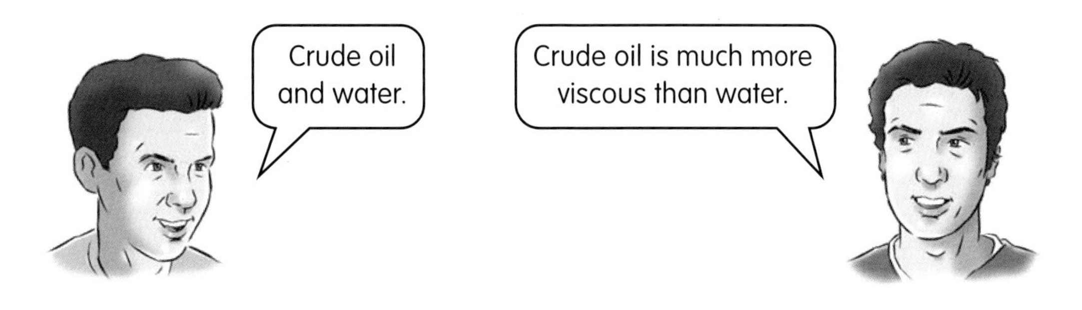

Lesson 3: Comparing more than two things

A **Work with a partner. Look at the table and discuss what you think the ticks and crosses mean. Where should you put ticks in the other columns?**

	Hard	Expensive	Elastic	Combustible	Versatile
glass	✓				
rubber	✗				
wood	✓				
steel	✓✓				
gold	✓				

B **Listen and complete the table above.**

C **Complete the sentences from the recording.**

1 Wood is ______________________ rubber, but steel is ______________________ substance.

2 Steel is ______________________ than wood, but gold is ______________________.

3 Rubber is ______________________________________.

4 Wood is ______________________________________.

D **Complete the rules.**

Superlative adjectives

When we compare more than two things, we use the superlative form of adjectives.

- Forming superlatives with one-syllable adjectives:

  the + adjective + ________

- Forming superlatives with longer adjectives:

  ________ + ________ + adjective

E Compare the different engines below. Use the questions as prompts.

1 Which engine is the most powerful?

2 Which one is the cheapest?

3 Which one do you think is the most useful?

4 Which one is the most sophisticated?

5 Which is the easiest to maintain?

6 Which is the most environmentally friendly?

7 Which is the most widely used?

8 Which one is the most dangerous?

F Complete the table. When do we use *less* and when do we use *the least*?

Electric motors are used Steam engines are used	more (<) less (>)	than	other engines.
____________ are ____________ are	more ____________ less ____________	than	combustion engines.

________ engines are ____________	the most the least	damaging to the environment.
____________ are ____________ are	the most the least	____________.

Lesson 4: Comparing metals

A **Read the statements about metals. Decide whether they are true (T) or false (F). Try to guess the meaning of the underlined words.**

1 Metals are conductive. They conduct electricity and heat well. ☐

2 It is possible to make wire out of most metals because they are ductile. ☐

3 Metallic substances are more brittle than nonmetallic substances. ☐

4 Pure metals are never soft or malleable. ☐

5 Durable metals like iron last a long time because they do not react with oxygen in the atmosphere. ☐

6 It is possible to blend metals together because they are fusible. ☐

7 Soft metals are sometimes mixed together to form 'alloys', which are harder and stronger. ☐

8 Shiny metals like steel and aluminium can be lustrous for many decades. ☐

B **Write an adjective for each quality in the table. Choose from the words in exercise A. Underline the suffix that creates the noun form.**

	Quality (noun)	Adjective
1	durability	______
2	ductility	______
3	conductivity	______
4	fusibility	______
5	hardness	______
6	brittleness	______
7	malleability	______
8	lustre	______

C **Listen and underline the stressed syllable in each noun in exercise B.**

D **Write a quality next to each definition.**

1 hardness______: This refers to the ability to withstand abrasion, surface dents or scratches.

2 ______: This is the attribute of the metal to blend with another substance under the influence of heat.

3 ______: This is the quality that gives the metal a shiny and polished appearance.

4 ________________: This is the ability of the material to resist stretching and return to its original length once the load is removed.

5 ________________: This is the ability of the shape of the material to be altered in any direction by compressive forces such as hammering, pressing, rolling or bending without breaking. These materials are highly plastic.

6 ________________: This is the ability of the material to be stretched into long, thin shapes, reducing its cross-sectional area. These materials must be highly plastic.

7 ________________: This is the opposite of 'toughness'. This kind of material will crack or break before it bends. It is not elastic and it is not ductile.

8 ________________: This refers to materials that have the ability to conduct heat or electricity.

9 ________________: This refers to the ability of the material to withstand the natural elements.

E Look at the table, which indicates some qualities of three different metals, and complete the sentences below. Put the adjectives in the correct comparative or superlative form.

	Hardness	Tensile strength	Thermal conductivity	Malleability
zinc	✓	✓	✓✓	✓
copper	✓✓	✓✓✓	✓✓✓	✓✓
iron	✓✓✓	✓	✓	✓✓

1 Copper is ____________________ (*hard*) zinc.

2 Iron is ____________________ (*hard*).

3 Zinc and iron have similar tensile strengths, but copper has ____________________ (*high*) tensile strength.

4 Iron has ____________________ (*thermal conductivity*) copper.

5 Zinc is ____________________ (*malleable*).

F Work in groups. Discuss what properties are required for the following and which ones are inappropriate. Explain your answers.

1 Chains

2 Storage tanks

3 Heating elements

G Listen to the conversation and check your answers.

Lesson 5: Talking about pigging

A Underline the word that is the odd one out in each row.

1 steel	rubber	plastic	seal	foam
2 fluid	plug	liquid	solution	water
3 friction	by-pass	lubrication	elevation	pressure
4 damaged	dry	worn	wide	debris

B Look at the diagram of a pig and answer the questions.

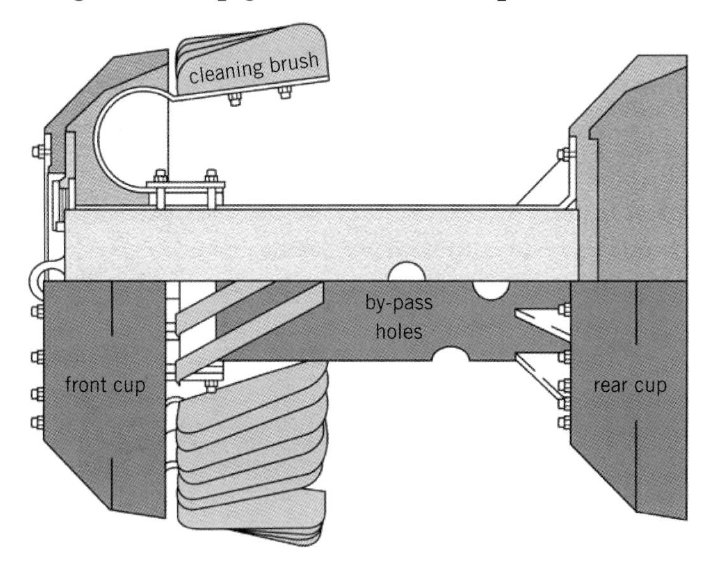

1 What does this device do?

2 Are these devices always the same size?

3 What is it made of?

4 Is it used in gas or oil pipelines?

C Read the texts about pigs and pigging and check your answers.

pig (*n*): a device that moves through the inside of a pipeline in order to clean or inspect or give information about the pipeline.

Pigging

Pigs (Pipeline Inspection Gauges) are pieces of equipment used to clean pipelines. The pig is propelled down the pipeline by compressed gas or fluid. Pigging systems and equipment are also used in smaller-diameter pipes in continuous and batch process plants.

PIPEFLOW SERVICES

cleans pipelines by propelling a series of flexible foam PRO-PIGS (cleaners) through them. Where there are heavy build-ups, the PRO-PIGS are introduced into the system in a progressive manner. We start with smaller and softer cleaning units, and gradually work up to full-sized cleaners with more abrasive qualities. This PRO-PIG cleaning method has proven to be the safest, fastest and most economical way to clean most pipelines.

D Read some more information about pigging. Decide whether the statements below are true (T) or false (F).

A pig generally has a steel body fitted with rubber or plastic cups at both ends. The cups are wider in diameter than the internal diameter of the pipeline to ensure a tight seal. Through use, the cups can become worn and need to be replaced, as pigs are less effective and allow more blow-by (fluid passing the pig) if the seal is not good.

Pigs are used to clear debris and excess liquid from the inside of pipelines. They are forced through the pipeline by the pressure of the flow. Several factors affect the force required to move the pig: elevation of the pipeline, friction and lubrication. A pig moves more easily through a crude oil pipeline than a dry gas pipeline due to better lubrication and greater flow pressure, but crude oil pipelines also contain much more debris than gas pipelines.

The pig contains by-pass holes which allow more fluid to flow past the pig, decreasing the speed. If all the by-pass holes are plugged, the pig moves more slowly, as the pressure forcing it through the pipeline is lower than when some of the by-pass holes are unplugged.

1 Pigs are usually made of foam or plastic. ☐

2 Pigs are very durable. ☐

3 It is possible to replace the rubber sections of the pig. ☐

4 They need an electric motor to propel them through the pipe. ☐

5 Pigs move faster through gas pipelines because there is less resistance. ☐

6 Pigs move faster through oil pipelines because there is higher pressure and less friction. ☐

7 Pigs do not move as fast when the by-pass holes are plugged. ☐

E Discuss what can go wrong with the pigging process. Use some of the words in the box.

corrosion	debris	blockage	worn	broken	cracked

Lesson 6: Expressing similarities

A **Sometimes when we compare things, we want to talk about similarities as well as differences. Look at the two oil drums. Decide whether the statements below are true (T) or false (F).**

A

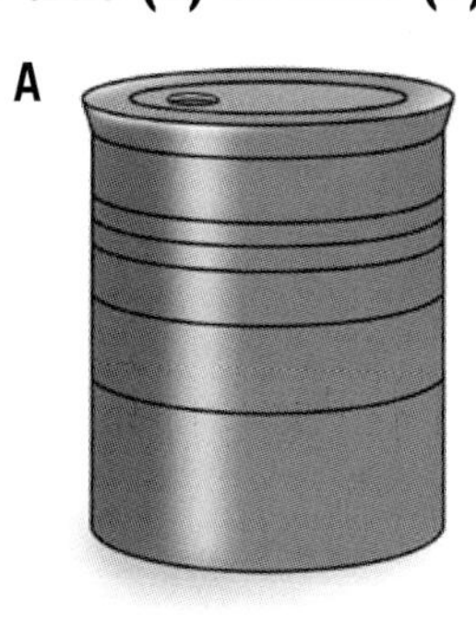

B

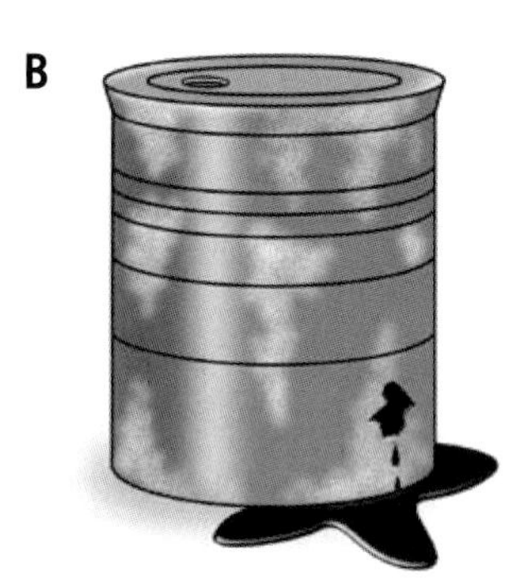

1 A is as big as B. ☐

2 B is not as big as A. ☐

3 Both A and B are probably made of steel. ☐

4 A is probably older than B. ☐

5 There is much less corrosion on A than on B. ☐

6 They are both grey. ☐

B **Which sentences focus on the similarities between A and B? Which sentences focus on the differences?**

C **Use the structures in exercise A to write:**
1 three sentences showing the similarities between water tanks C and D.
2 three sentences showing the differences.

C

D

1 C is ____.

2 Both ____.

3 They ____.

4 ______________________________

5 ______________________________

6 ______________________________

D Choose the correct option to complete each sentence.

1 Brass is not as *dense* / *density* as iron.

2 The main pipeline has a wider internal diameter *as* / *than* the relief pipe.

3 This pipe is almost as wide *as* / *than* the main pipeline.

4 *Both* / *Both of* oil and water are liquids.

5 The sump tank and the separated water tank *both are* / *are both* new.

E Compare the pigs in the picture below. Describe differences and similarities.

____

____

____

____

____

____

____

____

____

Lesson 7: Measuring temperature

A Look at the thermometers below and discuss the questions with a partner.

1 What materials are thermometers made of?

2 What different contexts are thermometers used in?

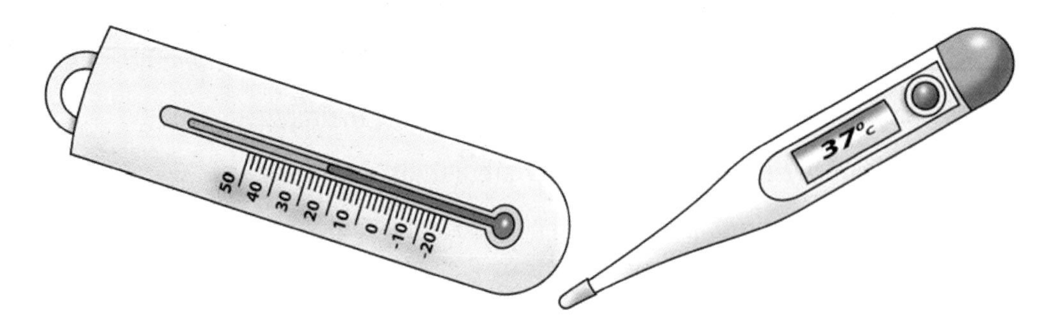

B Some thermometers contain mercury, some contain pentane and some contain alcohol. Listen to the comparison of thermometers and complete the table with the names of the three liquids.

Liquid	Range (in degrees Celsius)
	–80 to +70
	–35 to +510
	–200 to +30

C Complete the text with comparative and superlative forms.

Alcohol has the [1] __________ (*short*) range and mercury has the [2] __________ (*long*) range. Alcohol measures temperatures [3] __________ (*low*) than mercury, but [4] __________ (*low*) temperatures are measured using pentane. Mercury is used to measure [5] __________ (*high*) temperatures. Liquid in gas thermometers is [6] __________ (*weak*) than is necessary for industrial plant use, so industrial thermometers use bulbs and stems made out of steel, which is [7] __________ (*strong*) glass. These systems are completely filled with a liquid, gas or vaporizing liquid. Liquid-filled thermometers are [8] __________ (*popular*) gas or vaporizing alternatives because their range is [9] __________ (*long*).

There are also solid expansion-type thermometers. These work on the principle that different metals expand at different rates, i.e., brass expands [10]___________ (*quickly*) invar. One example of solid expansion-type thermometers is the bi-metal strip, where brass and invar are bonded together and fixed at one end. When the brass expands, the invar in the strip bends, showing the temperature on a scale, or triggering a switch.

D **Discuss what types of thermometers are shown below, and where they might be used.**

E **Write a comparison of the similarities and differences between the two types of thermometer.**

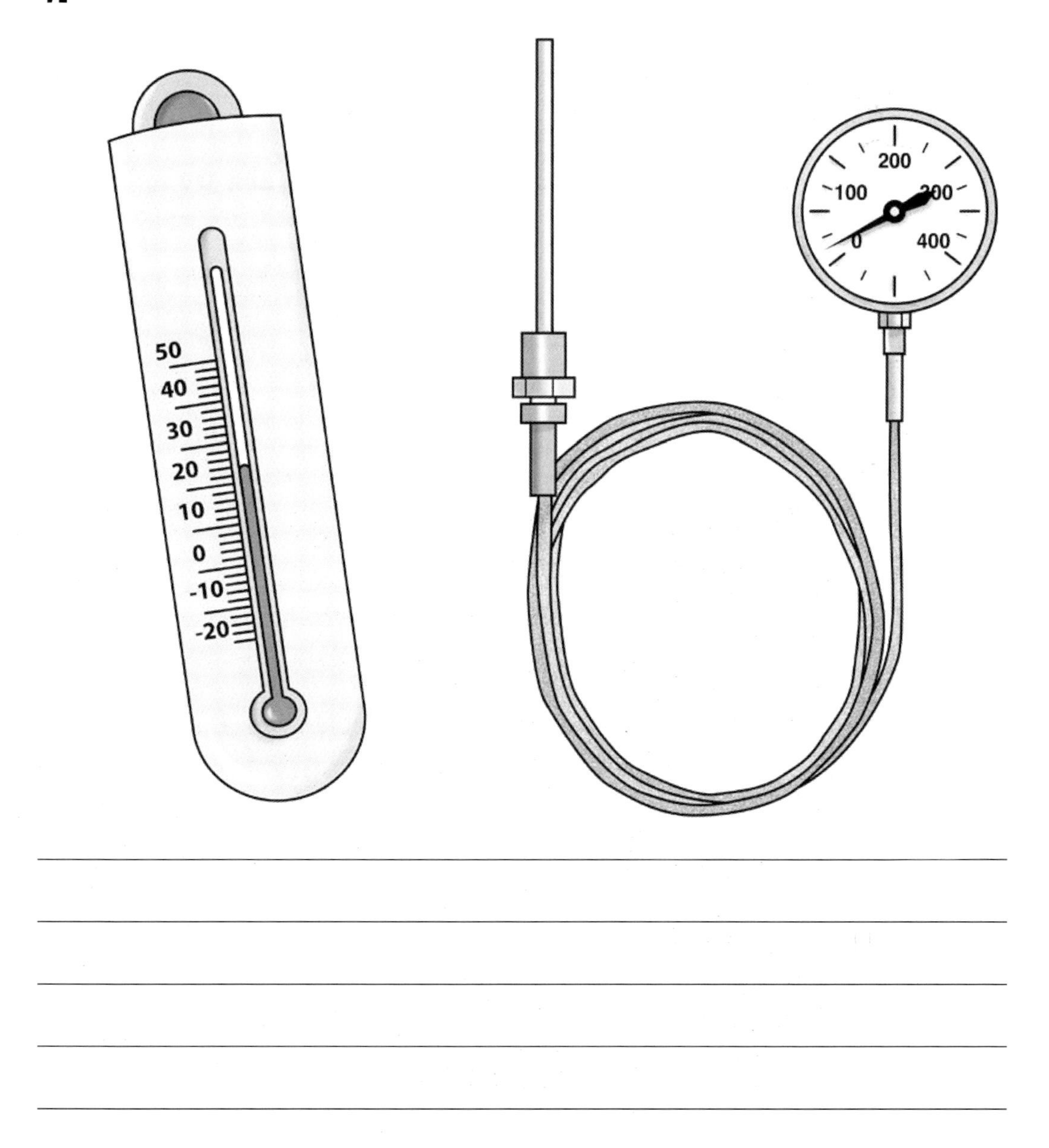

____

____

____

____

____

Lesson 8: Describing states of matter

Matter

Matter exists in three states: gas, liquid or solid.

A **Identify the three states in the picture below. Discuss what you know about each state.**

B **Decide which of the following are gases, which are liquids and which are solids.**

solvents mercury ice water vapour sugar solutes LPG crude oil grease

C **Listen to the mini-lecture about matter and try to complete the notes.**

Definitions

- Solids have definite volume and shape – strong forces hold molecules together.
- Liquids have definite volume, but not ________ – forces holding molecules together ____________________.
- Gases ____________________ – forces holding molecules together ____________________.

Changing states

- Matter changes from one state to another by ____________________ (molecules vibrate more than normal).
- Heat energy applied to a solid changes it to ________. Heat applied to liquid ____________________.

Combinations of elements

- All matter is made from elements = substances that cannot be ____________________.
- Compounds = substances where ____________________ and can only be separated ____________________.
- Mixture = ____________________ (can be separated by physical means).
- Solution = ____________________, or a solute (dissolved solid) and a solvent (liquid).

D **Answer the questions using full sentences. Check your answers with a partner.**

1 How many states of matter are there? What are they?

____

2 What are the properties of a solid?

____

3 Why does a solid change to a liquid?

____

4 What is the difference between a compound and a mixture?

____

5 What is the difference between a solute and a solvent?

____

E **Read the text and complete the table with the missing numbers. The numbers represent the amount of solute in grams. (Don't show your partner!)**

The rate a solute dissolves in a solvent is affected by particle size, temperature and agitation. Usually, more solute will be dissolved at higher temperatures than at lower temperatures. One of the exceptions to this is calcium sulphate, where the amount of solute dissolved decreases above 40°C.

By raising the temperature, we can make a supersaturated solution which contains solute dissolved at a higher level than the solubility limit. A supersaturated solution is more unstable than a normal solution as it contains too much of the solute.

2.10 2,040 370 2.03 354 366 2,600 2.11

	10°C	20°C	30°C	40°C	50°C	60°C
sodium chloride		360	363			374
calcium sulphate	1.90	2.05			2.08	
sugar	1,910		2,200	2,380		2,870

F **Work with a partner. Ask each other questions to check your answers.**

Lesson 9: Talking about oil products

A What different products are made from petroleum? Complete the diagram with the words in the box.

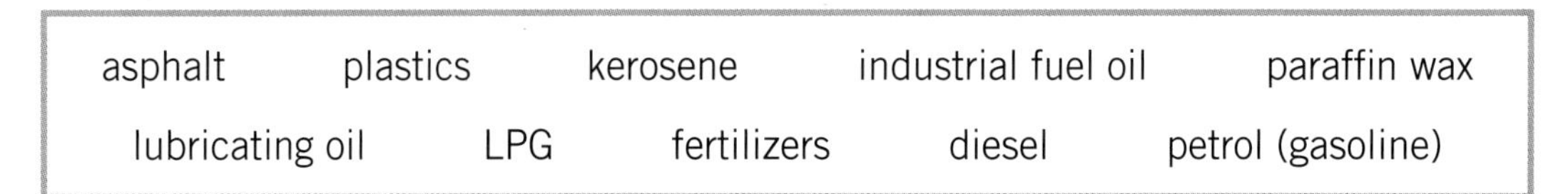

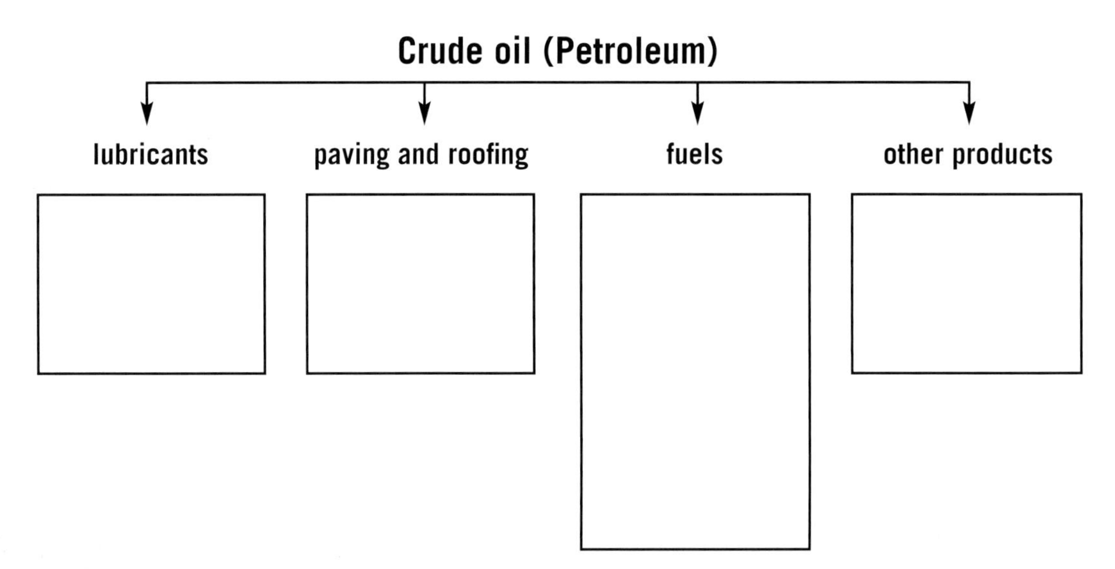

B Read the text and check your answers.

Crude oil, or petroleum, is a thick, black mixture of hundreds of different compounds, and is practically useless when taken from the ground. About 88% is processed and refined into fuel, asphalt and lubricants. The remaining 12% is converted into other materials such as plastics, and chemical products such as fertilizers and solvents.

Several types of motor, aviation and heating fuel are produced in oil refineries. These are: LPG (liquefied petroleum gas), petrol (gasoline), kerosene and diesel.

Kerosene and LPG are more versatile and are used in domestic heating systems and portable stoves. The primary use of kerosene, however, is as aviation fuel in jet engines. LPG is starting to be used as a 'green gas' to power cars because it produces fewer emissions than either petrol or diesel fuels.

Petrol and diesel are the most widely used fuels for cars, buses and trucks. Most cars have petrol engines, and larger vehicles, e.g., buses, tend to have diesel engines. Diesel engines operate at much higher pressures than petrol engines and, therefore, they have to be built more strongly. However, diesel engines are more efficient than petrol engines and are more economical – they can use up to 40% less fuel.

Diesel is used in industry alongside heavier industrial fuel oil. This is used in ships and factories, and is normally burned in a furnace.

In addition to motor fuels, important products refined from oil are lubricants. Lubricants are essential to ensure moving parts in a machine work smoothly and stop them overheating. Different lubricants are suitable for different jobs. Engine oil is less viscous. Grease is much thicker and is used to protect sealed bearings. Multigrade lubricants can operate at different temperatures without being affected. Lubricants without a temperature-sensitive polymer are more viscous at low temperatures than at high temperatures.

C **Complete the table below with information from the text.**

Oil product	Uses	Properties
petrol/gasoline		
diesel		
LPG		
fuel oil		
engine oil		
grease		
multigrade lubricants		

D **Look at the diagram showing the products made from a typical 42-gallon barrel of crude oil. Discuss what it shows and what you think the missing parts of the key are.**

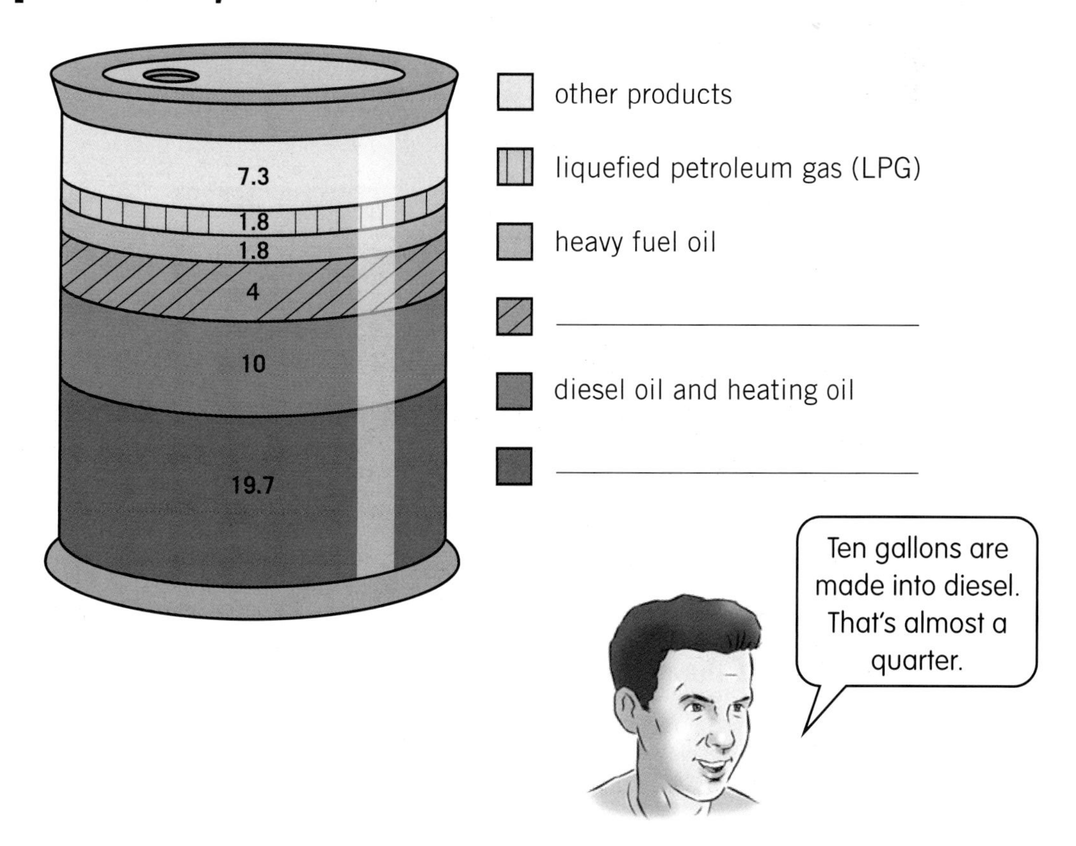

Lesson 10: Talking about fractional distillation

A Work in groups. Discuss the following questions.

1 Why is oil a valuable commodity?

2 What is the most important use of oil?

3 How is crude oil changed into different products?

4 Why do some countries export crude oil and not separate it into different products?

5 Are there better energy sources than oil?

B Look at the diagram below. It shows the process of fractional distillation, which separates petroleum into its different components. In your groups, discuss what happens. Use the words in the box.

heat	boil	temperature	vapour	rise	column	trays	bubble caps	condense

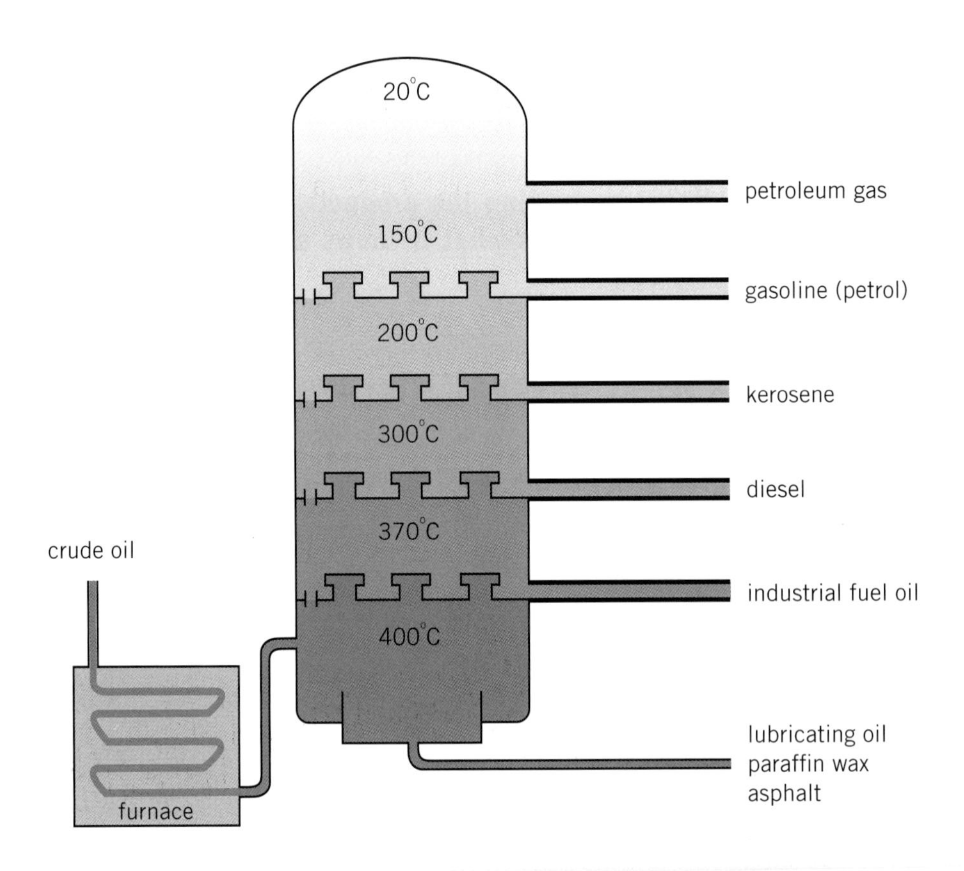

C Listen to someone explaining the fractional distillation process. Which of the questions in exercise A does he answer?

D **Listen again and decide whether the statements below are true (T) or false (F).**

1 It is easy to separate the different components of gasoline. ☐

2 All components have the same boiling temperature. ☐

3 The column is the same temperature at the top and the bottom. ☐

4 There are trays inside the distillation column. ☐

5 Substances with high boiling points condense at the top of the column. ☐

6 LPG and diesel condense at the top of the column. ☐

7 The collected liquid always needs further chemical processing. ☐

8 The gasoline yield from each barrel can be increased by chemical processing. ☐

E **Reorder the description of the first part of the distillation process below. Complete the description in your own words.**

☐ When the mixture boils, it forms vapour.

☐ The vapour rises in the column.

☐ The petroleum is heated to a high temperature (about 600° Celsius) in a steam boiler or furnace.

☐ The vapour enters the bottom of a long column that is hotter at the bottom (400–600° Celsius) and cooler at the top (about 20° Celsius). It has trays inside it with holes or bubble caps in that allow vapour to rise through them.

5 ______________________________

____

____

6 ______________________________

____

____

7 ______________________________

____

____

UNIT 7

Review: Making comparisons

A Answer the questions below as quickly as possible. Check your answers with a partner. If you disagree, discuss your reasons.

1 Does Russia produce more or less oil than Saudi Arabia?

2 Is diesel more or less flammable than petrol?

3 Which is the most versatile: gold, titanium or platinum?

4 Offshore oil is slightly more difficult to extract than onshore oil. True or false?

5 Which is the most combustible: kerosene, butane or fuel oil?

6 Does America produce as much oil as it consumes?

7 Is natural gas more or less difficult to transport than crude oil?

8 Both petroleum and gasoline produce asphalt. True or false?

9 What are the main differences between methane and propane?

10 What are the similarities between oil and gas transportation?

B Write five more questions using structures (e.g., comparatives and superlatives) from the unit.

1 ______________________________

2 ______________________________

3 ______________________________

4 ______________________________

5 ______________________________

C Work with a partner. Think about what types of lubricants there are, and why lubrication is important.

D Work with a partner. Read about different types of lubrication.
Student 1: Look at page 221.
Student 2: Look at page 231.

Swap information about the advantages of each type, then discuss and write down some disadvantages.

UNIT 7

Assess your skills: Making comparisons

Complete (✓) the tables to assess your skills.

I can ...	Difficult	Okay	Easy
• compare two things (objects, devices or systems).			
• compare more than two things.			
• explain similarities and differences between objects such as thermometers.			
• describe the qualities of metals.			
• understand and describe the pigging process.			
• listen to information about states of matter and take notes.			
• understand texts about oil refining and oil products.			
• talk and write about the oil-refining process.			

I understand ...	Difficult	Okay	Easy
• the unit grammar: comparative adjectives			
adverbs of degree			
superlative adjectives			
as ... as			
grammar patterns with *both*			
• the unit vocabulary (see the glossary)			

If there is anything you are not sure of, ask your trainer to revise the material.

UNIT 8

DESCRIBING PROCESSES AND PROCEDURES

The aim of this unit is to:

- provide the structures necessary to describe processes logically and coherently

By the end of this unit, you will be able to:

- describe events using sequencers
- describe basic processes
- identify different word forms
- discuss oil separation
- describe past experiences

Lesson 1: Sequencing simple processes

A **Look at the instructions below. Decide what they are for. Then complete them with the words from the box in the correct form. You can use some of the words more than once.**

voltage	replace	loose	tight
final	ensure	cover	secure

How to ________________

1 First, ________________ the device is power-isolated.

2 Second, use a screwdriver to remove the holding screws from the plug and remove the ________________.

3 Third, check that the screws holding the wires in place are ________________.

4 Next, ________________ the screws which hold the fuse in place and remove the fuse.

5 ________________ it with a new fuse of the same ________________.

6 Then, ________________ the screws which hold it in place.

7 After that, ________________ the cover and ________________ it with the holding screws.

8 ________________, check the device now works.

B **Read the description below and rewrite it in sentence form, using similar numbered instructions and sequencers.**

use a wrench to loosen the wheel nuts use a jack to raise the car remove the wheel nuts remove the wheel put the new wheel in place replace the wheel nuts and lower the car tighten the wheel nuts

How to change a wheel

1 ______________________________

2 ______________________________

3 ______________________________

4 ______________________________

5 ______________________________

6 ______________________________

7 ______________________________

8 ______________________________

wheel nut

jack

C **Work in pairs. Discuss the order in which you would put on your PPE. Sequence the actions below using *First*, *Second*, *Third*, *Next*, *After that*, etc.**

put your arms in the overalls	put on your safety boots	put your legs in the overalls	put on your safety gloves
put on your ear protectors	fasten the overalls	put on your safety glasses	put on your hard hat

D **Listen and check your answers.**

E **Look at the pictures below. Discuss what you need to consider when you lift a heavy load. Include some of the prompts in the boxes.**

Before you lift the load, …

While you are holding the load, …

When you lift the load, …

After you move the load, …

1

2

3

4

assess …

put your feet …

make sure your back …

keep your hands …

Lesson 2: Talking about safety procedures and electricity

A **Find ten words in the wordsearch that are connected with injuries and accidents. Use them to complete the sentences below.**

A	R	T	D	S	P	W	C	D	G	H	K
M	F	H	E	O	R	E	T	E	C	A	U
R	I	P	L	A	S	H	O	C	K	R	N
O	R	S	E	A	N	E	R	A	U	E	C
U	S	X	T	I	N	G	U	S	S	C	O
M	T	A	I	D	I	B	T	U	H	O	N
S	A	F	E	T	Y	N	W	A	J	V	S
A	I	X	B	R	M	P	U	L	S	E	C
B	D	S	P	E	L	W	B	T	N	R	I
S	S	H	P	A	E	N	T	Y	M	O	O
U	I	A	N	T	I	D	C	R	H	N	U
T	O	L	O	M	A	I	B	U	R	N	S
A	N	L	N	E	M	O	N	I	T	O	R
D	I	A	G	N	O	S	E	T	I	C	E
B	D	W	T	T	N	Z	T	S	L	K	R
P	Z	M	N	Y	E	X	Q	L	K	N	I
F	W	E	T	F	K	C	O	A	N	B	U
G	N	I	R	Q	D	Y	M	S	O	P	F

1 Tomas had an electric s _ _ _ _ and suffered minor b _ _ _ _.

2 Bob went on a f _ _ _ _ - a _ _ course.

3 Have you had any s _ _ _ _ _ training?

4 The c _ _ _ _ _ _ _ _ was u _ _ _ _ _ _ _ _ _ _ _ and his p _ _ _ _ was weak.

5 The t _ _ _ _ _ _ _ _ helped Ahmed r _ _ _ _ _ _.

6 A first-aider should assess the situation and d _ _ _ _ _ _ _ the problem.

B **Note down your answers to the questions and compare your ideas in groups of three or four.**

1 What sort of dangers and accidents are associated with electricity?

____

____

____

2 What do the pictures show?

____

____

3 How can you ensure there are no accidents when working on electrical equipment?

____

____

C Match each correct verb in column A with the phrase in column B to outline the procedure for energy isolation.

A	B
1 Inform	the maintenance work or inspection.
2 Turn off	the main disconnect switch.
3 Turn off	the isolation.
4 Lock	the point of operation of the device.
5 Apply	the main disconnect switch.
6 Test	all parties of the work to be done.
7 Conduct	a warning tag to the main disconnect switch.

D Listen and check your answers. What is the name of the procedure?

E Complete each sentence with the correct time word in the box (sometimes more than one is possible). Then answer the questions.

before	when	while	during	after

The main disconnect switch must be turned off [1]________ an inspection.

[2]________ you are working on the equipment, the power must be turned off.

You should inform everyone involved [3]________ you turn off the power.

The power can be restored [4]________ the locks and tags are removed.

1 Which time word is often used with continuous tenses?

2 Which time word always goes in front of a noun or noun phrase?

3 Which time word can often replace *while*?

Lesson 3: Giving first-aid

A Choose the verb in each row that goes best with the words in bold.

1 preserve	diagnose	examine	**life**
2 provide	assess	call	**for help**
3 give	promote	preserve	**recovery**
4 provide	diagnose	give	**the problem**
5 prevent	examine	give	**treatment**
6 assess	diagnose	promote	**the situation**

B Work with a partner. Discuss these questions. Use some of the phrases from exercise A.

1 What is first-aid?

2 What does a first-aider do?

C Listen and make notes.

The aims of first-aid:

____

____

____

Responsibilities of a first-aider:

____

____

D **Write instructions for what you should do if you see an accident. Listen again and check you have the correct sequence.**

1st ____

2nd ____

3rd ____

4th ____

5th ____

6th ____

E **Four of the sentences below have mistakes in them. Identify which four sentences are incorrect and rewrite them with the correct time expression.**

1 Vasily did a first-aid training course three months ago.

2 There was an accident in the workshop the last week.

3 This was the first time Vasily used first-aid since he completed the training.

4 Vasily looked after the casualty during fifteen minutes.

5 The medical team arrived ten minutes later.

6 Later in afternoon, Vasily wrote an accident report.

7 The next day, he visited the hospital.

8 Day before yesterday, the casualty left hospital.

____

____

____

____

F **Work with a partner. Talk about your experiences of accidents or first-aid. Use some of the expressions in the box.**

Last year, … Last month, … Several years ago, … Later, …
The first time … The day I … The last time … The day before …
The next day, … I … for … I … during …

Lesson 4: Using the ABC rule

A **Complete the table with the correct body parts.**

ribs throat forehead chest chin lips lungs heart

Parts of the head and neck area				
Parts of the torso				

B **Discuss what is happening in the pictures. Label the pictures with the verbs in the box.**

kneel blow compress check tilt pinch

1
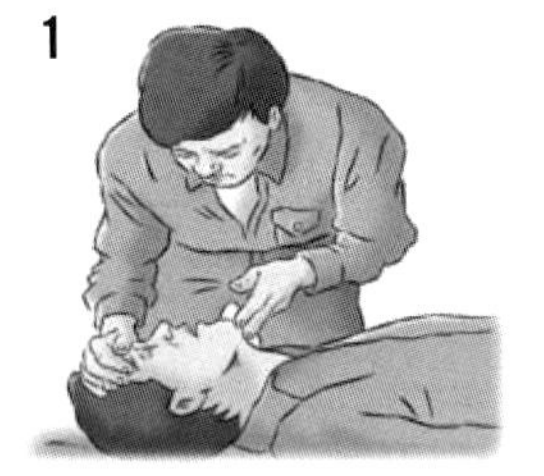

2
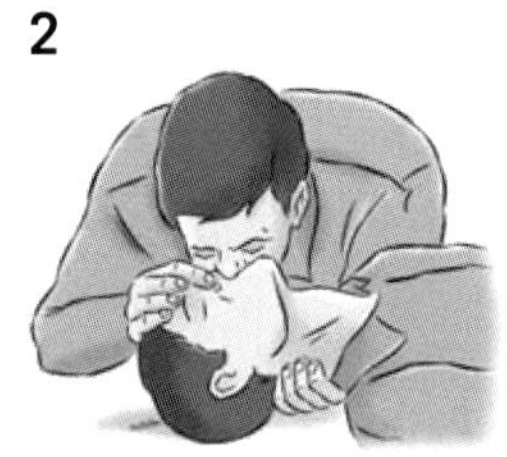

3
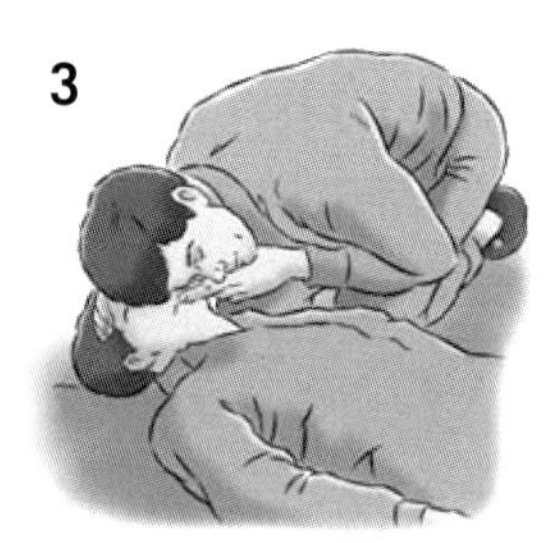

4
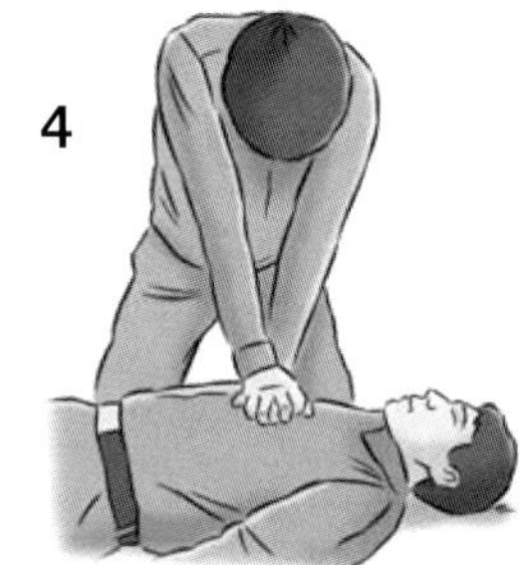

C **Read the ABC rule and check if you were right.**

If the casualty is not breathing, and if the heart is not beating, it is essential that you maintain circulation and respiration until trained medical help arrives. This process follows the ABC rule.

A is for airway

The casualty's throat may be blocked. Check to make sure the airway is open. If the airway is blocked, it is vital that you clear it so oxygen can reach the lungs. Open and clear the airway immediately by lifting the chin forward and tilting the forehead back. When the airway is open, the casualty may begin breathing for themselves. If not, ...

B is for breathing

Breathing for a casualty is called artificial ventilation. This normally involves blowing gently into the mouth of the casualty. (Sometimes this is not possible if they have serious facial injuries, are face down or if there is evidence of a corrosive substance around the mouth.) Before blowing into the casualty's mouth, pinch their nose, then open your mouth and put your lips around the mouth. Watch the casualty's chest for movement during the artificial ventilation. This indicates the lungs are being filled, or that the casualty is breathing again naturally.

C is for circulation

After you have given artificial ventilation for a few breaths, you should check whether the casualty's blood is circulating by checking their pulse. If there is no heartbeat, the oxygenated blood will not be circulating. To help start the heart pumping, you should start external chest compressions immediately. The ratio of compressions to breaths is 15:2. It is very important to place the hands in the correct position (two finger widths above the bottom of the junction of the ribs) and to keep your arms straight while you are doing it. Apply the compressions regularly, but you should stop as soon as you see a spontaneous pulse in the casualty's neck. Check again after a minute and then at three-minute intervals.

D Put the procedures below in the correct order (A–L).

1 Place your hands on the chest. ☐
2 Seal your lips around the mouth. ☐
3 Check whether the casualty is breathing. ☐
4 Take a deep breath. ☐
5 Watch the chest rise and fall. ☐
6 Check the pulse. ☐
7 Open your mouth wide. ☐
8 Apply compressions. ☐
9 See if the casualty's throat is blocked. ☐
10 Blow into the casualty's mouth. ☐
11 Pinch the nose closed. ☐
12 Lift the chin and tilt the forehead back. ☐

E Read the sentences and underline the mistakes. Then rewrite them correctly.

1 I have worked offshore in 2001. ________________

2 Have you ever see an explosion? ________________

3 I have started my shift at 7.00 a.m. ________________

4 I haven't work offshore. ________________

5 You have been abroad? ________________

6 I haven't never been in an accident. ________________

Lesson 5: Describing past experiences

A **Listen to Bob being interviewed about his experience with first-aid training. Decide whether the statements below are true (T) or false (F).**

1 Bob did a safety training course a year ago. ☐

2 He studied fire protection during the course. ☐

3 He has not used his first-aid training. ☐

4 He helped put out a fire last month. ☐

B **Look at the sentences below from the recording and answer the questions.**

I did a safety training course about three years ago.

I have used the fire safety training.

1 Which sentence refers to a definite time in the past?

____

2 Which sentence refers to an unspecific time in the past?

____

C **Match the timelines with the correct rules about the use of the past simple and the present perfect.**

Past simple and present perfect

The past simple gives information about a specific time in the past, e.g., *I studied first-aid in 1997.*

Past — Now — Future

1977

The present perfect gives general information about the past. We do NOT know when the action happened, e.g., *I have studied first-aid.*

Past — ? ? — Now — Future

D **Complete the table below to show the form of the present perfect.**

+	*I*	________		*done*	*first-aid training.*
	Subject	Auxiliary verb *have/has*		Past participle	
-	*He*	________	*not*	________	*first-aid training.*
	Subject	Auxiliary verb *have/has*	*not*	Past participle	
?	________	*you*	________		*first-aid training?*
	Auxiliary verb *Have/Has*	Subject	Past participle		

E **Write the past simple and past participle of the verbs in the table below. Then listen and check your answers.**

Infinitive	Past simple	Past participle
visit	visited	visited
use		
work		
study		
do		
go		
fly		
see		

F **Complete the sentences below with the verbs from the table above in the present perfect.**

1 Ahmed ________________ a first-aid course.

2 Frank ________________ offshore and he ________________ in a helicopter many times.

3 Tom ________________ a blowtorch before.

4 ________ you ever ________________ electronics?

5 I ________________ to Qatar, but I ________________ Kuwait.

6 I ________________ never ________________ an accident.

Lesson 6: Talking about events that have/have not happened

A What sort of things have you learned to do in your job? Read the questions in the questionnaire, then add two more. Interview another student and complete the table.

General information | **Specific information**

Have you ever ...		*When ...?*	*Where ...?*	*...?*
... done safety training?	Yes			
	No			
... studied first-aid?	Yes			
	No			
... seen an accident?	Yes			
	No			
... been injured at work?	Yes			
	No			
... worked offshore?	Yes			
	No			
... flown in a helicopter or plane?	Yes			
	No			
	Yes			
______________________?	No			
	Yes			
______________________?	No			

B **Tell the group about one of the things your partner has done. Give as much detail as possible.**

C **Look at these two sentences and answer the questions below. Then check your answers with a partner.**

1 Which technician did a first-aid course before now?

2 Do we know when he did it?

3 Are either of them doing a course now?

4 Will Khalid do a first-aid course in the future?

5 Do we know when?

6 When do we use *already*?

7 When do we use *yet*?

Khalid

Mohammed

D **Listen and complete Vasily's maintenance checklist. Discuss with a partner which things he has already done and which things he hasn't done yet.**

Maintenance checklist	
1 Identify the problem.	✓
2 Choose the best solution.	
3 Complete a risk assessment.	
4 Inform people about the job.	
5 Select the correct equipment.	
6 Isolate the power.	
7 Conduct the maintenance.	
8 Test the system.	
9 Inform people the job is complete.	
10 Reconnect the power.	

E **Write about an activity, course or project that you have recently taken part in. Say what you have done and not done yet.**

Lesson 7: Talking about job interviews

A **Identify the documents in the pictures. Which of the following can you see?**

a CV   a job ad   a contract   a certificate

CURRICULUM VITAE

Name: Ahmed Abdulkader

Born: 8th September, 1972
Oran, Algeria

Nationality: Algerian

Our client, a leading Oil & Gas service company, has an opportunity for a Dive Representative to join their team. You will spend approximately one week in an office, then three weeks offshore, overseeing the diving operations. There may be additional work after the project is complete.

Job Type: Contract

Location: Aberdeen / North Sea
UNITED KINGDOM

B **Ahmed is applying for the job in the advert above. Read the job description. What questions do you think he will be asked in the interview?**

C **Listen and tick the questions that you hear. What are Ahmed's answers?**

1 How long have you worked in the oil industry? ☐
2 What positions have you had? ☐
3 When were you in Azerbaijan? ☐
4 Have you worked offshore? ☐
5 Have you done any in-service training courses? ☐
6 When did you do your training? ☐
7 Where did you do your training? ☐
8 How long have you studied English? ☐
9 Why do you want to leave your current job? ☐

D **Look at the sentences and timelines. What is the difference between ...**

1 the concepts?

2 the tenses?

Ahmed worked in Kuwait for two years.

Past — Now — Future

Ahmed has worked in Azerbaijan for two years.

Past — Now — Future

2 years ago

E **Complete each phrase with *since* or *for*.**

1 ______ ages

2 ______ forty minutes

3 ______ I was born

4 ______ last year

5 ______ three years

6 ______ March

7 ______ two months

8 ______ 1993

9 ______ I started work

10 ______ about an hour

11 ______ two months ago

12 ______ the company began

F **Complete the rules.**

since/for

- We use ______________ when we want to talk about periods of time.
- We use ______________ when we want to say when the action started.

G **You are going to role-play a job interview. Work with a partner. Interview your partner for a job in their field.**

1 Write down six questions to ask at the interview.

2 Conduct the interview and note down your partner's answers.

3 Swap roles.

____

____

____

____

____

Lesson 8: Talking about heating and thermostats

A Try to answer the questions about heat and thermostats. Compare your answers with a partner.

1 Conduction occurs when the atoms in a material increase their vibration due to heat. The heat passes through the material. Which substances are good conductors of heat?

____

____

2 How is convection different from conduction?

____

____

3 What is radiation?

____

____

4 What happens to a liquid when it reaches its boiling point?

____

____

5 What factors affect the boiling point of a substance?

____

____

6 What is the difference between a thermometer and a thermostat?

____

____

B Read the information about thermostats, then answer the questions.

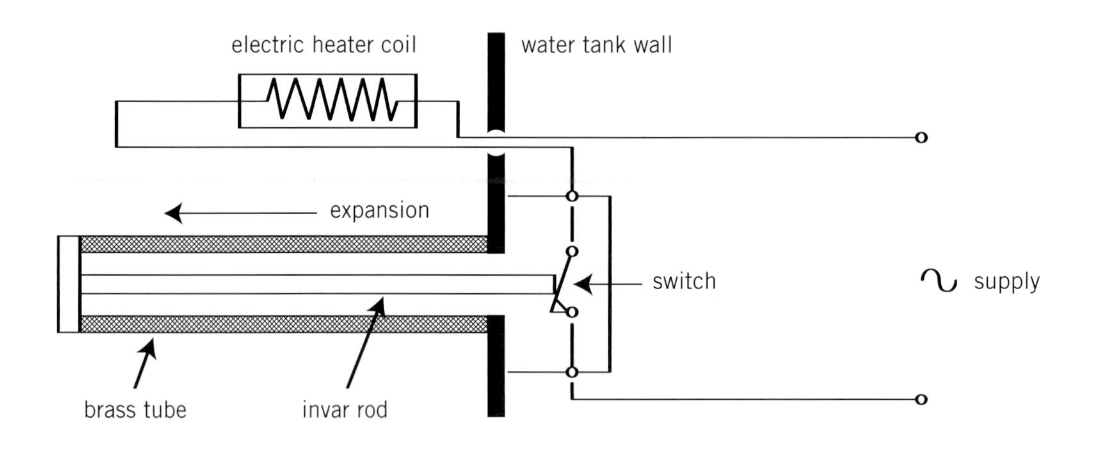

Thermostats are a cheap method of regulating temperature. Although they are not very accurate, they can maintain a constant temperature ± 3°C.

The diagram on page 170 shows a simple rod thermostat. When the liquid in the tank is cool, the switch on the thermostat is in a closed position. This completes the circuit so the electric heater heats the water.

When the water reaches the required (set) temperature, the brass tube gets hot and expands a lot, whereas the invar rod does not expand. The expanded brass tube pulls the invar rod away from the switch. This opens the switch and breaks the circuit so the electric heater is disconnected. The electric heater will stay disconnected until the brass tube contracts enough to close the switch again. The set temperature can be adjusted by changing the tension of the spring which closes the switch.

1 Why are invar and brass used in the system?

____

____

2 How is the circuit broken?

____

____

3 How is the circuit made?

____

____

4 How is the set temperature adjusted?

____

____

C Look at the diagram of a domestic heating system and identify the following parts.

the hot water storage tank	the boiler	the cold water supply tank

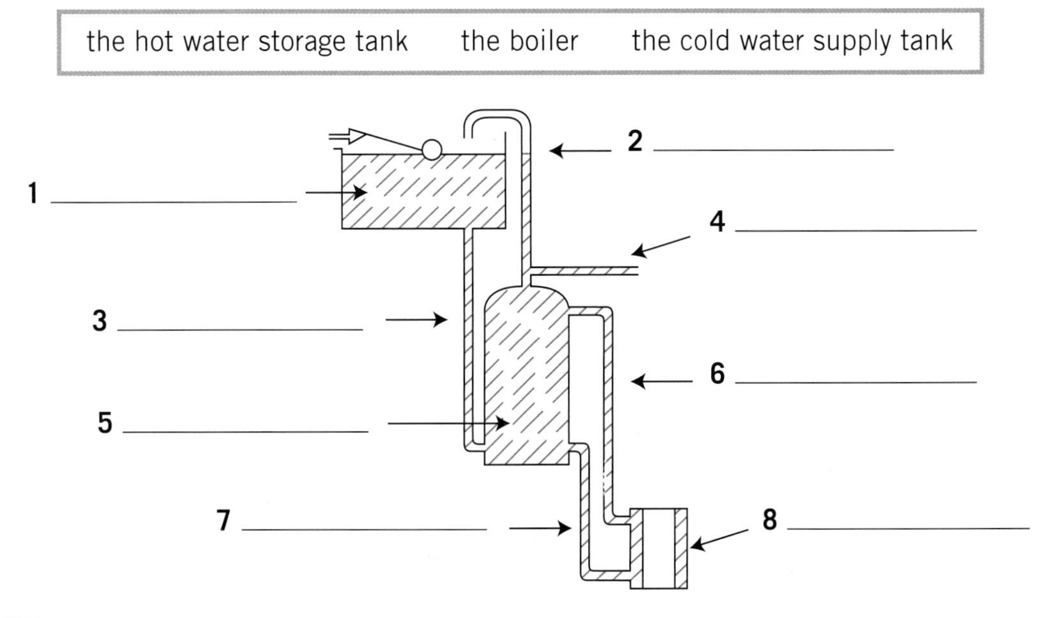

D Listen and finish labelling the system.

Lesson 9: Talking about oil and water separation

A Complete the text with the words in the box.

emulsion	chemical	crude	gas	water	heat	oil

[1]__________ oil taken from the ground is an emulsion (mixture) of oil, [2]__________ and gas. The water and [3]__________ need to be separated from the [4]__________.

This can be done by using either:

[5]__________ treatments: these work by changing the density and viscosity of the __________ OR

[6]__________ treatments: these use demulsifiers which react with the oil and water emulsion and break it down chemically.

B Work with a partner. Look at the diagram of a heater treater below and discuss how it works. Use the questions on page 173 to help you.

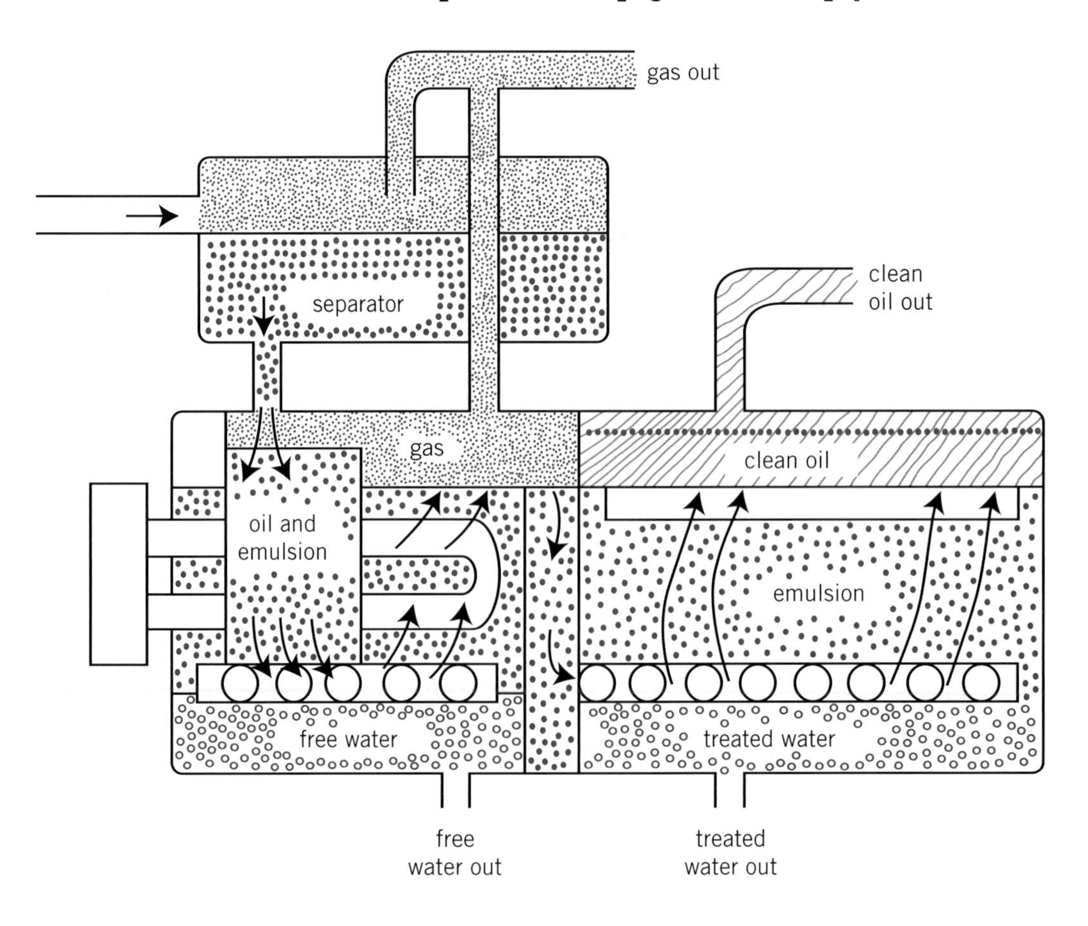

1 What is the first thing removed from the emulsion?

2 Is water removed before heating?

3 At what stages of the treatment is gas removed?

4 Which is heavier, oil or water?

C Complete the notes about how the process works.

1 The emulsion is piped into the separator and ____________________.

2 After ________________ has been removed, the ________________ is heated and ____________________.

3 Then, the emulsion ____________________.

D Look at the diagram of a chemical separator and expand the notes into sentences. Be careful to use the correct prepositions.

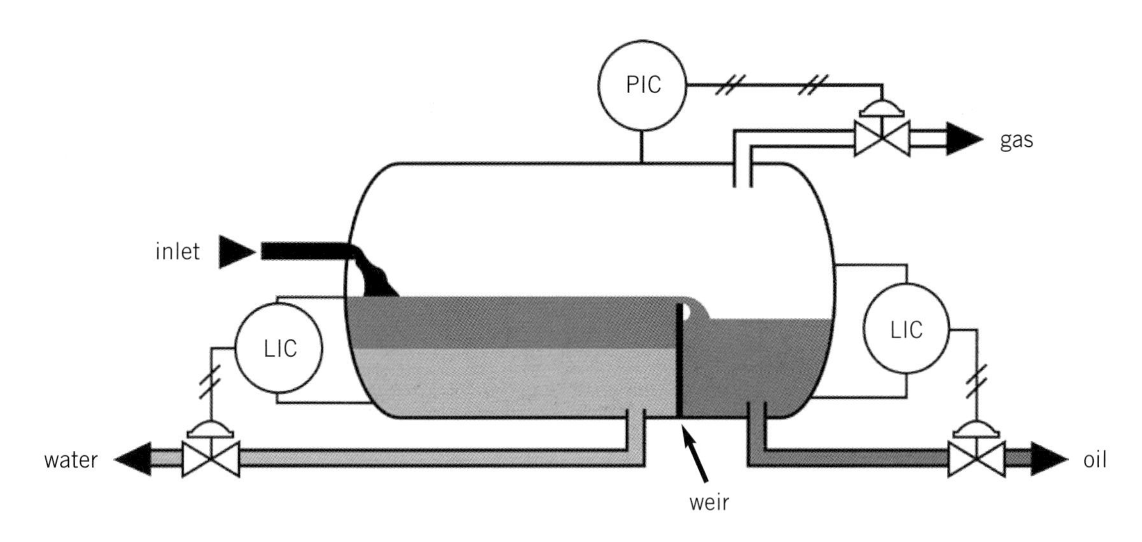

This is a diagram of a horizontal-type production separator. It is designed to remove water from oil. It works with the addition of a chemical demulsifying agent.

1 flows / the inlet valve / left-hand side of the weir

2 sinks / the bottom of the tank / flows / the release pipe

3 floats / the surface / cross the weir / flows / the release pipe

4 pressure / maintained / vent / gas / relief pipe / and / adjust / relief valves / pipes

E Listen and compare the description with yours.

Lesson 10: Describing the final stages of separation

After the oil has been separated from the water using heat or chemical treatments, the final stages of separation involve using a centrifuge.

A **Work with a partner.**
Student 1: Look at page 222.
Student 2: Look at page 232.

Read the description of how a centrifuge works and label as much of the diagram below as you can. Then swap information with your partner and finish labelling the diagram.

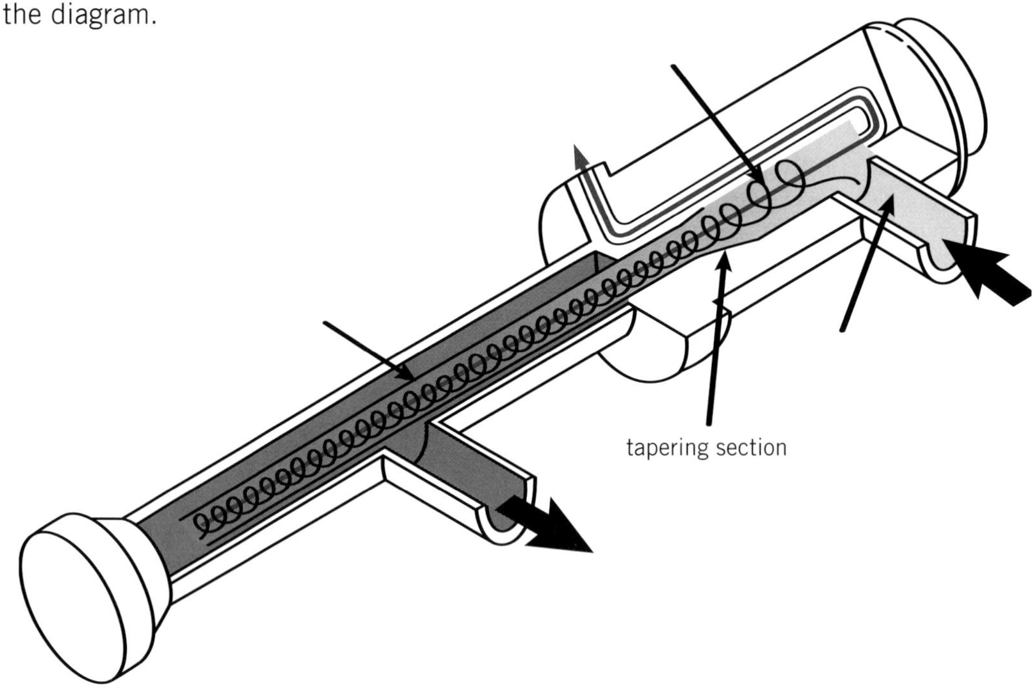

B **Prepare a short talk about how a centrifuge works. Use the words in the box.**

force	heavier	outer wall	flow
reject	central core	base	dense

____

____

____

____

____

C **Look at the diagram below. See if you can identify the tag numbers and abbreviations. Compare your ideas in small groups.**

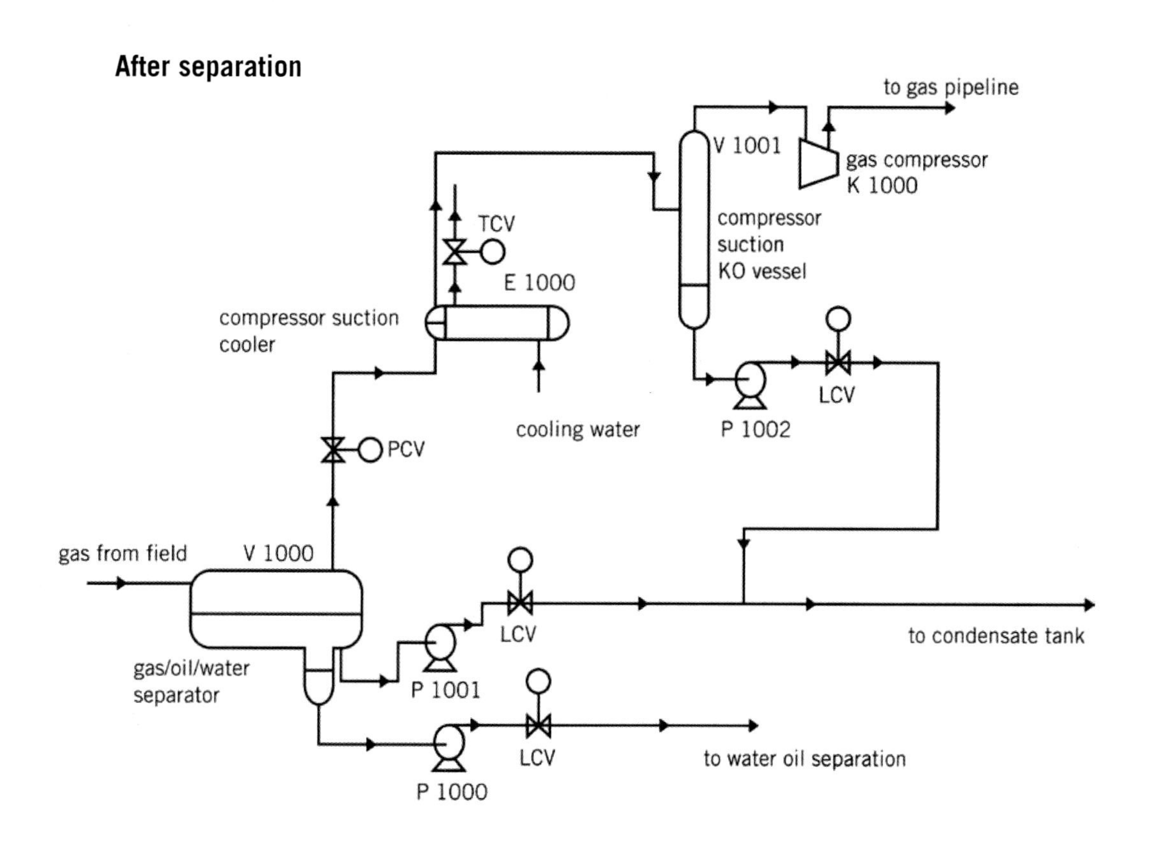

1 V1000 = ______________________

2 PCV = ______________________

3 P1001 = ______________________

4 LCV = ______________________

5 E1000 = ______________________

6 KO = ______________________

D **In your groups, prepare a short presentation of how the process works.**

____

____

____

____

____

____

UNIT 8

Review: Describing processes and procedures

A Complete the table with words from the unit.

Noun	Verb	Adjective
		boiling
compression		
		demulsified
	isolate	
separator		
		reconnected
recovery		
	conduct	
		unconscious
	treat	
		centrifugal
breath		

B Divide the nouns into three groups.

1 Nouns connected with first-aid and injuries:

____

2 Nouns connected with heat and electricity:

____

3 Nouns connected with processes for separating oil:

____

C Complete the text with verbs in the correct tense (present simple, past simple or present perfect).

During the last 25 years, there [1]____________ (*be*) improvements in safety in the oil industry. Nevertheless, there [2]____________ (*be*) still some major accidents. In March 2001, explosions [3]____________ (*kill*) 25 people at the world's biggest offshore rig in Brazil, which shows that the oil industry [4]____________ (*be*) not completely safe today. Many companies [5]____________ (*increase*) the number of safety training programs for managers and employees over the past three or four years. They [6]_____ also _____ (*start*) to try out new ideas, such as last-minute risk assessments and the use of green hats for new or inexperienced workers.

UNIT 8

Assess your skills: Describing processes and procedures

Complete (✓) the tables to assess your skills.

I can ...	Difficult	Okay	Easy
• describe sequences.			
• use appropriate linking words to show the time relationship between events.			
• give information about first-aid and the process of artificial respiration.			
• describe ongoing actions/situations.			
• use appropriate structures to describe completed and unfinished experiences.			
• ask questions and give information suitable for a job interview.			
• explain how a thermostat works.			
• describe diagrams that show oil and water separation processes.			

I understand ...	Difficult	Okay	Easy
• the unit grammar: sequencers			
present perfect			
present perfect vs past simple			
for/since			
• the unit vocabulary (see the glossary)			

If there is anything you are not sure of, ask your trainer to revise the material.

UNIT 9 GIVING ADVICE

The aims of this unit are to:

- practise giving advice
- discuss real and unreal situations using the language of hypothesis

By the end of this unit, you will be able to:

- use appropriate modal verbs to express obligation
- give advice in the active or passive voice
- make deductions and discuss possibilities about problems
- give a short toolbox talk offering advice and information
- speculate about hypothetical and unreal situations
- use formal and informal technical expressions appropriately

Lesson 1: Talking about transportation

A Identify the vehicles, machines and people in the pictures below.

B Complete the table on page 179 with the words in the box. Think of more words to add to each category.

pedestrian	ABS	registration document
seat belt	driving licence	passenger

People who are involved with vehicles	Documentation	Safety features of vehicles

C Read the text, which gives advice about safety. Decide whether the statements below are true (T) or false (F). Try to guess the meaning of the underlined words. Then compare your answers with a partner.

> A driver has certain responsibilities to ensure his own safety, the safety of the vehicle and the safety of any passengers, pedestrians and other road users.
>
> Larger vehicles have more safety regulations because they carry more people. In addition to the minimum requirements of all road-ready vehicles, larger vehicles should also be fitted with airbags, ABS, side-impact bars and air conditioning. Regardless of size, all vehicles must be fitted with a first-aid kit, fire extinguisher and a warning triangle for use in an emergency.
>
> Before using any vehicle, the driver must check it is roadworthy and fit for purpose. These checks include the relevant documentation, e.g., the registration certificate and insurance certificate. If possible, the driver should also examine the maintenance record of the vehicle.

1 It is necessary to have insurance to drive. ☐

2 It is mandatory to have the maintenance record of the vehicle to drive. ☐

3 All vehicles have to have side-impact bars. ☐

4 The driver must check if the vehicle is appropriate for the job. ☐

5 Air conditioning is optional in larger vehicles. ☐

6 The driver shouldn't worry about other vehicles on the road. ☐

7 Smaller vehicles do not require a first-aid kit. ☐

8 All vehicles need to carry a first-aid kit. ☐

D Read the text again and add words from the text to the table in exercise B.

E Work with a partner. Look at the statements in exercise C. Think of a different way to say each statement.

Lesson 2: Talking about rules

A Read the statements about vehicles. Are they all correct?

1 You must check that any vehicle is roadworthy before you drive it.

2 You mustn't drink alcohol the night before you pilot a helicopter.

3 You have to keep to the speed limit.

4 You don't have to wear a seat belt to operate a forklift.

5 Tanker drivers had to work longer hours in the past.

6 Women couldn't pilot planes in World War II.

7 Passengers didn't have to wear seat belts until 1990 in England.

B Choose the correct modal verb to complete each rule.

must* and *have to

Must and *have to* are modal verbs that express obligation. They are very similar in meaning. Both can often be used in affirmative sentences.

- *Have to* / *Must* tends to be used when we talk about personal obligation, i.e., things we feel are necessary to do.
- *Have to* / *Must* tends to be used when we talk about obligations that are imposed on us, i.e., things that an outside authority tells us are necessary to do.
- *Don't have to* / *Mustn't* is used to talk about negative obligations, i.e., things we are not allowed to do.
- *Don't have to* / *Mustn't* is used when there is no obligation, i.e., we can choose to do it, or not.
- Only *have to* / *must* can be used in the past tense.
- Negative obligation in the past is expressed using *mustn't* / *couldn't*.

C Complete the table below.

Present	Necessary	Unnecessary
Affirmative	must ________________	
Negative	________________	don't have to

Past	Necessary	Unnecessary
Affirmative	________	
Negative	couldn't	________

D **Choose the correct modal verb to complete each sentence.**

1 You *mustn't / don't have to* wear PPE at home.

2 You *mustn't / don't have to* distract a driver when they are driving.

3 You *mustn't / don't have to* ignore safety signs.

4 You *mustn't / don't have to* use another language during English classes.

5 You *mustn't / don't have to* get up early when you are on holiday.

E **Complete the sentences with the correct modal form.**

1 Jim ________ start work at 7.30 tomorrow. He has an urgent job to do.

2 I need more exercise. I ________ get up early to go for a run before work.

3 The doctor told me I ________ stop drinking so much alcohol.

4 You ________ cut an electric cable with a pair of scissors.

5 This job isn't urgent. You ________ start it until after lunch.

6 I ________ use my mobile phone when I was in the meeting.

7 Do passengers ________ wear seat belts?

8 When I was on the rig, we ________ wear shorts.

F **Complete the sentences with your own ideas. Compare your sentences with a partner.**

1 Drivers of large vehicles have to ________________.

2 When you are a pedestrian, you don't have to ________________.

3 Passengers mustn't ________________.

4 Pilots must be good at ________________.

5 In the past, oil workers had to ________________.

Lesson 3: Comparing advice and obligation

A **A supervisor is talking to Jim, a new roustabout who is starting work on the floor of the rig. Listen to the conversation and answer the questions.**

1 What will Jim have to do at first?

2 What does he have to wear?

3 What does the supervisor advise Jim to do if he has any questions?

4 What other advice does he give?

5 What does the supervisor say Jim mustn't do?

B **What is the difference between the sentences below? When do we use *should* and *shouldn't*?**

You must read the safety regulation handbook today.

You should read the safety regulation handbook today, if possible.

You mustn't smoke on the floor of the rig.

You shouldn't smoke on the floor of the rig.

C **Read the sentences and underline the mistakes. Then rewrite them correctly.**

1 If you feel tired, you mustn't go to bed earlier.

____

2 You don't have to obey the safety regulations.

____

3 Do you think I must do this welding now?

____

4 I think you must ask for time off when your wife has her baby.

____

5 We must wear a life jacket when the sea was rough.

____

D Complete the sentences with *must*, *have to*, *mustn't*, *should* or *can*.

1 You ____________ drive the truck if the oil warning light comes on.

2 Pedestrians ____________ walk across the rig.

3 There is a delay because the crane driver ____________ wait for the signal person.

4 If you have any problems, you ____________ speak to your supervisor.

5 You ____________ clear up this rubbish. It's your job.

E What advice would you give to the people below? Discuss your ideas with a partner and write an answer to each problem.

1 'I saw a colleague taking some tools home from the workshop.'

____

____

____

2 'I'm very tired by the end of my 12-hour shift.'

____

____

____

3 'I think there are too many safety rules in my job.'

NO SMOKING

HARD HAT AREA

GLOVES MUST BE WORN

____

____

____

4 'My job is boring.'

____

____

____

F Listen and check your answers.

Lesson 4: Troubleshooting

A **Work in groups. Look at the pictures and predict what is going to happen. Listen to some people discussing each picture and see if their ideas are the same as yours.**

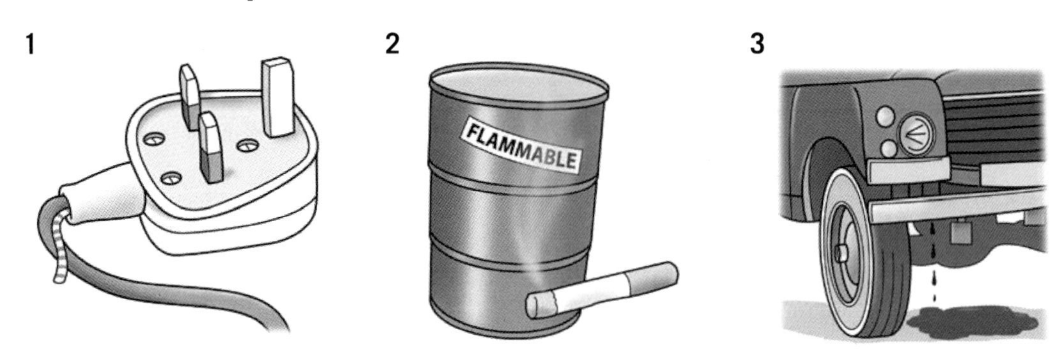

B **Listen again and complete the sentences in the box below.**

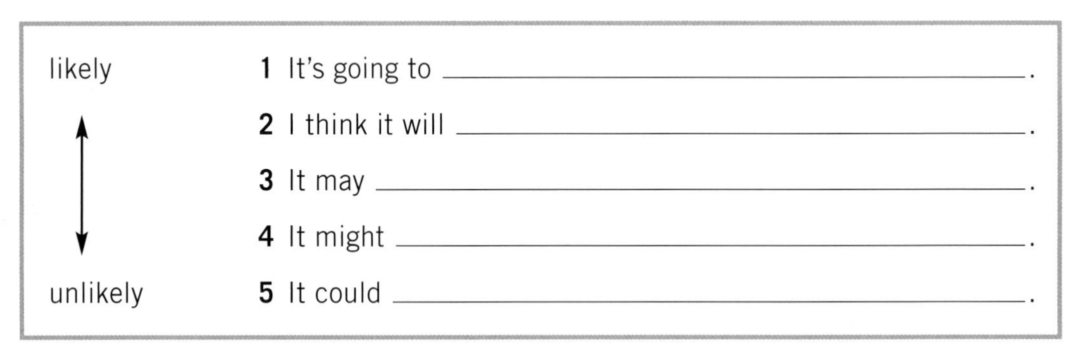

likely	**1** It's going to ________________.
↑	**2** I think it will ________________.
	3 It may ________________.
↓	**4** It might ________________.
unlikely	**5** It could ________________.

C **Look at the situations below. What might/may/could be the cause? Suggest a cause for each and a consequence.**

1 A tank is overflowing.

____

____

2 A light is not working.

____

____

3 Water is not flowing from the tap.

____

____

4 An operator has a headache.

____

____

D Compare the two sentences using modals of possibility.

It might injure someone.

The bulb might have blown.

1 Which one refers to the past?

2 When is a modal of possibility followed by the infinitive of the verb?

3 When is it followed by the 3rd-person form of the verb (*have* + past participle)?

E It is important to be able to anticipate problems. Look at the pictures below. Explain what the problem is and what might/may/could happen.

1

____

____

____

2

____

____

____

3

____

____

____

It is important to consider all the evidence when you look for the cause of a problem. You must also be able to identify what is relevant and what is irrelevant.

Lesson 5: Talking about hazards

A What are the slip hazards in the pictures?

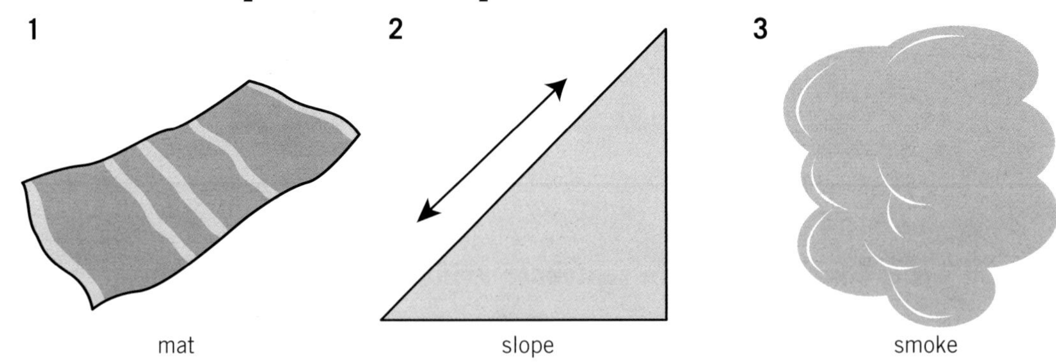

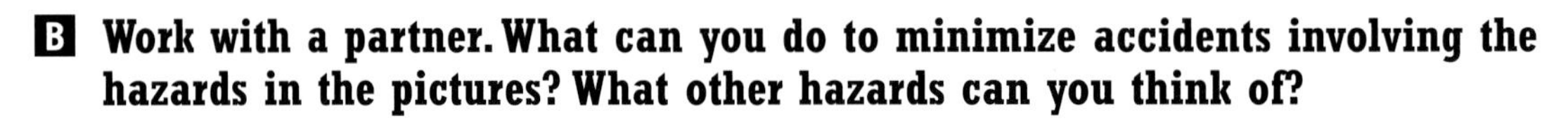

B Work with a partner. What can you do to minimize accidents involving the hazards in the pictures? What other hazards can you think of?

C Match each rule and piece of advice (a–h) with the correct hazard (1–8).

Hazard	Rules and advice	Hazard	Rules and advice
1 spillage of wet or dry substances	*h*	5 lighting	
2 trailing cables		6 steps and level changes	
3 miscellaneous rubbish		7 slopes	
4 rugs/mats		8 steam or smoke obscuring view	

Rules and advice	
a Equipment must be positioned to avoid cables crossing pedestrian routes. If possible, use cordless tools.	**e** This should be avoided. If it cannot be avoided, you should improve lighting and add high, visible tread indicators.
b You should ensure mats are securely fixed and do not have curling edges.	**f** It is a good idea to provide handrails and use floor markings.
c You should try to improve lighting levels to ensure more even lighting.	**g** Areas have to be kept clear and rubbish should be removed.
d This should be eliminated or controlled by redirecting it away from risk areas and by improving ventilation. Employees must also be warned of it.	**h** These have to be cleaned up immediately; if a liquid is greasy, make sure a suitable cleaning agent is used. You should use appropriate barriers to tell people the floor is still wet.

D **Look at exercise C. Underline all the sentences with modal verbs in the active voice. Circle all the sentences with modal verbs in the passive voice.**

E **Rearrange the words to make passive sentences.**

1 by / faulty / injuries / may / caused / equipment / be / .

____

2 all / times / safety / must / followed / be / at / regulations / .

____

3 job / be / appropriate / always / for / tool / should / used / the / the / .

____

4 investigation / has / completed / to / be / an / incident / an / after / .

____

F **Complete the sentences in a logical way. Use either active or passive sentences. Then compare your ideas with a partner.**

	should	______________________.
	shouldn't	______________________.
PPE	must	______________________.
	mustn't	______________________.
	can	______________________.
	can't	______________________.
	should	______________________.
	shouldn't	______________________.
First-aid	must	______________________.
	mustn't	______________________.
	might	______________________.
	might not	______________________.

What did you say about PPE?

I said PPE should always be worn in the workshop.

Lesson 6: Giving toolbox talks

A **Have you ever used any of these items? Match each name with the correct picture.**

1 fire extinguisher •

2 fire blanket •

3 fire bucket •

Toolbox talks

A toolbox talk is a short talk that is given to clarify what or how equipment should be used, what procedures should be followed or how a job should be done.

B **Read the example toolbox talk below about fire extinguishers, then choose the bullet point that best summarizes the main aim of the talk.**

- To ensure everybody knows what procedures to follow when there is a fire.
- To highlight the dangers of fires.
- To give information about fire equipment.

Firstly, if there is a fire, be sure someone has gone to call the immediate supervisor. Then, you should attempt to put the fire out with a fire extinguisher or other fire equipment.

There are several types of fire extinguisher and they should be used for different types of fires. Technicians must be able to identify what type of extinguisher should be used on different types of fires.

One basic type of extinguisher contains water.
- These should be used on fires of wood, paper, cloth and rubbish.
- They should not be used on burning liquids like gasoline. The burning liquid will float on the water and the fire could spread.
- They should also not be used on live electrical equipment because water is a good conductor of electricity and the firefighter could get an electric shock.

The other common types of extinguishers contain carbon dioxide or dry chemicals.
- These are usually mounted on or beside machinery, equipment and containers of flammable liquid.
- Carbon dioxide and dry chemicals cut off the air supply necessary for a fire and smother it.

C **Answer the questions.**

1 What is the purpose of the toolbox talk?

____

2 What do you think is the most important information in the talk?

____

3 Does the talk contain enough information? What information would it be useful to add?

____

D **Listen. What is the topic of the talk?**

E **Work with a partner. Prepare a toolbox talk on one of the following areas. Make notes in the table below.**

Working with electricity	Working at heights	Using power tools

Aim of the talk	
Point 1	
Point 2	
Point 3	

F **Give your talk to another pair or group.**

Lesson 7: Using formal and informal technical expressions

A **John is a reporter visiting the rig where Bob works. Bob is telling him how the machinery works. Listen and finish labelling the diagram below.**

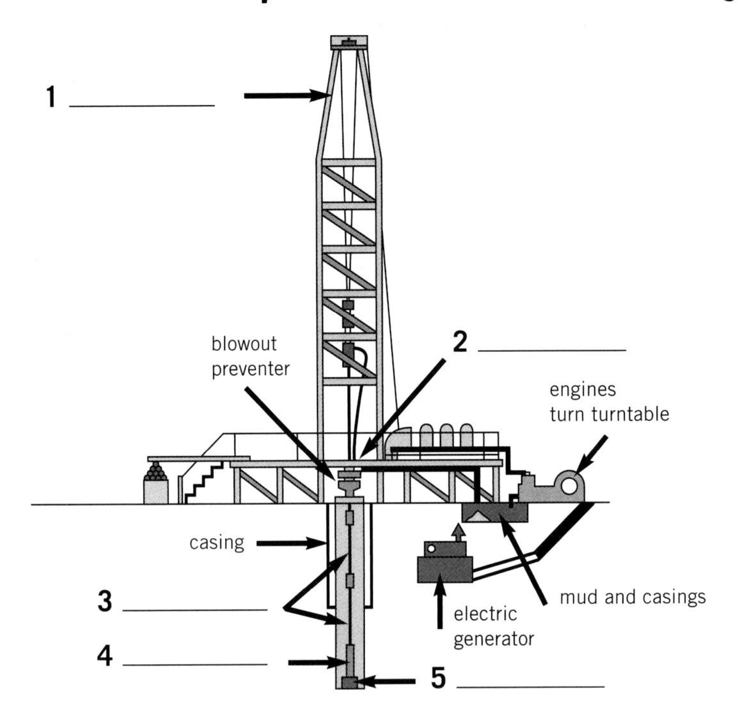

B **Read the next part of the dialogue and underline the expressions that John does not understand.**

Bob: This bit is dull, so we'll probably have to trip the pipe.

John: What does that mean?

Bob: The drill is probably worn out, so we'll have to take the pipe out of the hole to have a look at it.

John: And do you have any other problems?

Bob: Sometimes equipment gets lost in the borehole, then we have to fish for it.

John: Fish?

Bob: Things that are lost in the hole are called *fish* or *junk*.
If we lost a drill bit, for instance, we would send down a junk basket.

John: Is that a tool?

Bob: Yes, for hoisting the fish out of the hole. That involves making a round trip.

John: What's a round trip?

Bob: Pulling the drill string out of the well, then running it back into the well.
We only do it if we must. It's expensive and involves too much downtime.

John: It involves what?

C Read the text below. Underline the formal technical expressions and circle the informal technical expressions.

Recently, deeper oil wells have been constructed as it is becoming more difficult to locate hydrocarbon accumulations near the surface. At greater depths, there is a greater risk of a blowout. This is when a well blows out of control. It is known as a gusher. The mud weight must be carefully regulated to prevent this. Ground barium may be utilized to add weight to the mud.

If the drill bit strikes a high-pressure rock formation, however, the mud column may be insufficient to restrain the high-pressure gas, oil or water that will emerge as a kick. Unless the BOP rams can be closed quickly, this may result in a blowout.

In the case of a gas well blowout, it may be necessary to divert the gas into a flare pit, where it is ignited in order to avoid an explosion.

D Match each formal technical term in column A with its standard English term in column B.

	A	B
1	ignite	control
2	restrain	build
3	strike	use
4	prevent	hit
5	construct	find
6	risk	not enough
7	regulate	hold back
8	locate	stop
9	insufficient	set fire to
10	utilize	danger

E Complete the table with the phrases in the box according to whether they are formal or informal expressions.

It was ignited. The rig is in operation. We set it ablaze. Can you fix it?
He handled the situation. The rig is onstream. Can you repair it?
He kept the situation under control.

Formal expressions	Informal expressions

Lesson 8: Speaking hypothetically

A Work in small groups. Discuss the questions below.

1 What happens to oil when oil wells get older?

2 What can be added to the well to maximize recovery?

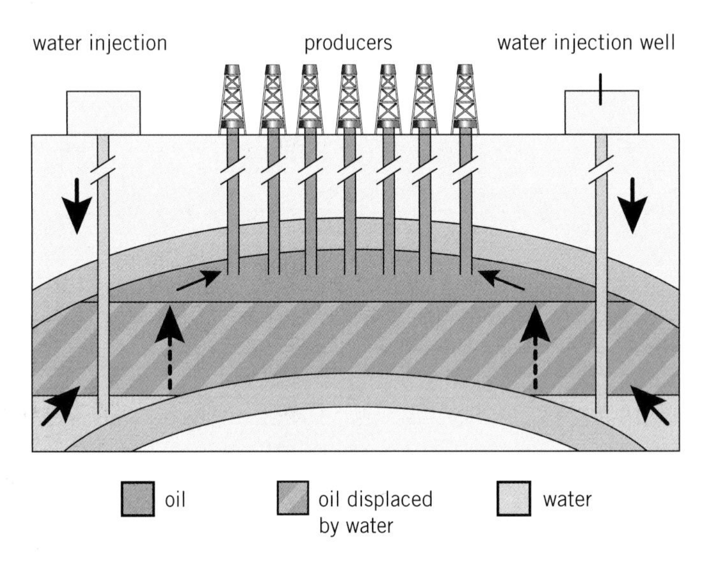

B Listen to Bob explaining about older oil wells. Compare his answers with yours.

C Listen again and decide whether the statements below are true (T) or false (F).

1 If we inject water into a well, oil recovery can improve by up to 15%. ☐

2 If water is injected into an oil well, it will raise the pressure of the well. ☐

3 If the pressure in a well was too low, there would be a blowout. ☐

4 If we didn't inject wells with water, they would produce less gas. ☐

5 If we used untreated seawater, it would be highly corrosive. ☐

6 If water is injected without being treated first, it can damage the equipment. ☐

D Look at the sentences in exercise C. Which sentences use the first conditional? What structure do the other sentences use?

E Complete the examples and the grammar rules with the words in the box.

be	past	possible	unreal	present	inject	modal	improve	would	used

1 First conditional: *If we* ________ *water into a well, oil recovery can* ________.

2 We use the first conditional to talk about ________ situations. It is formed using *if* + a ________ tense and *will* or a ________ verb + infinitive or base form.

3 Second conditional: *If we* ________ *untreated seawater, it would* ________ *highly corrosive.*

4 We use the second conditional to talk about hypothetical or ________ situations. It is formed using *if* + a ________ tense and ________ or a modal verb + infinitive or base form.

F Complete the sentences below with *will*, *won't*, *would* or *wouldn't*.

1 If you press that button, the machine ________ shut down.

2 I ________ earn more money if I worked offshore.

3 We ________ have to reassess the site for digging if it rains.

4 If we send our employees on a training course, we ________ need to hire new workers.

5 If I didn't work in the oil industry, I ________ work in telecommunications.

6 If there was no oil in Algeria, oil companies ________ operate there.

G Read the sentences below and decide whether each one is possible (P) or unlikely/impossible (U).

1 My English will improve next year. ☐

2 I will visit England next summer. ☐

3 This lesson will finish on time. ☐

4 I will earn more money. ☐

5 I will lose my job. ☐

6 The oil reserves in this country will run out in the next 20 years. ☐

Lesson 9: Talking about drilling mud circulation systems

A Match the words to make a phrase or term connected with the drilling mud circulation system.

1 rock	shaker
2 drill	trap
3 rotary	man
4 shale	bit
5 mud	cuttings
6 sand	table

B Complete the text with the phrases and terms from exercise A.

Drilling mud circulation system

The circulation system pumps drilling mud. It is pumped to the bottom of the well, where it cools the ________________ and returns to the surface bearing ________________ from the drilling. This ensures that the drill bit drills into uncut formation rather than redrilling old cuttings.

The 'mud' is actually a mixture of clay, weighting material and chemicals. It is pumped through the kelly, the ________________, the drill pipes and the drill string under high pressure.

After the drill mud has returned to the surface through the annulus, the rock cuttings are separated from the mud in the ________________. The mud then passes through a ________________ which removes smaller solid particles. It is then recycled through a series of mud tanks and is finally pumped through a centrifugal mixing pump to the mud-mixing hopper, where new mud is mixed.

The mud system is controlled by an engineer known as the ________________. He tells the floormen how to mix the mud, and analyzes the content of the mud if there is a problem.

C **Work with a partner. Look at the diagram below and take it in turns to ask questions about the function of different parts of the system.**

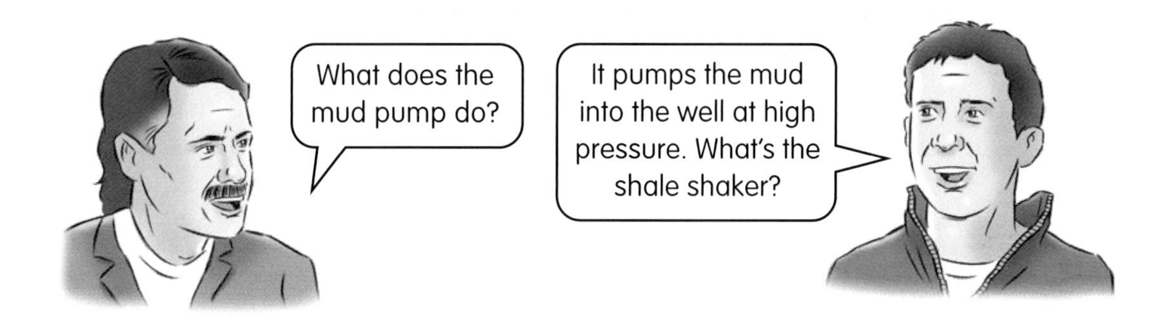

D **Work in pairs or small groups. Discuss your ideas about the questions below. Use the first or second conditional.**

1 What will happen if the drill bit is worn out?

2 What would happen if the pressure in the well decreased?

3 What would happen if the mud was not heavy enough to support the sides of the borehole?

4 What would happen if the drill pipe became blocked?

5 What would happen if a cutter from the drill bit got lost in the well?

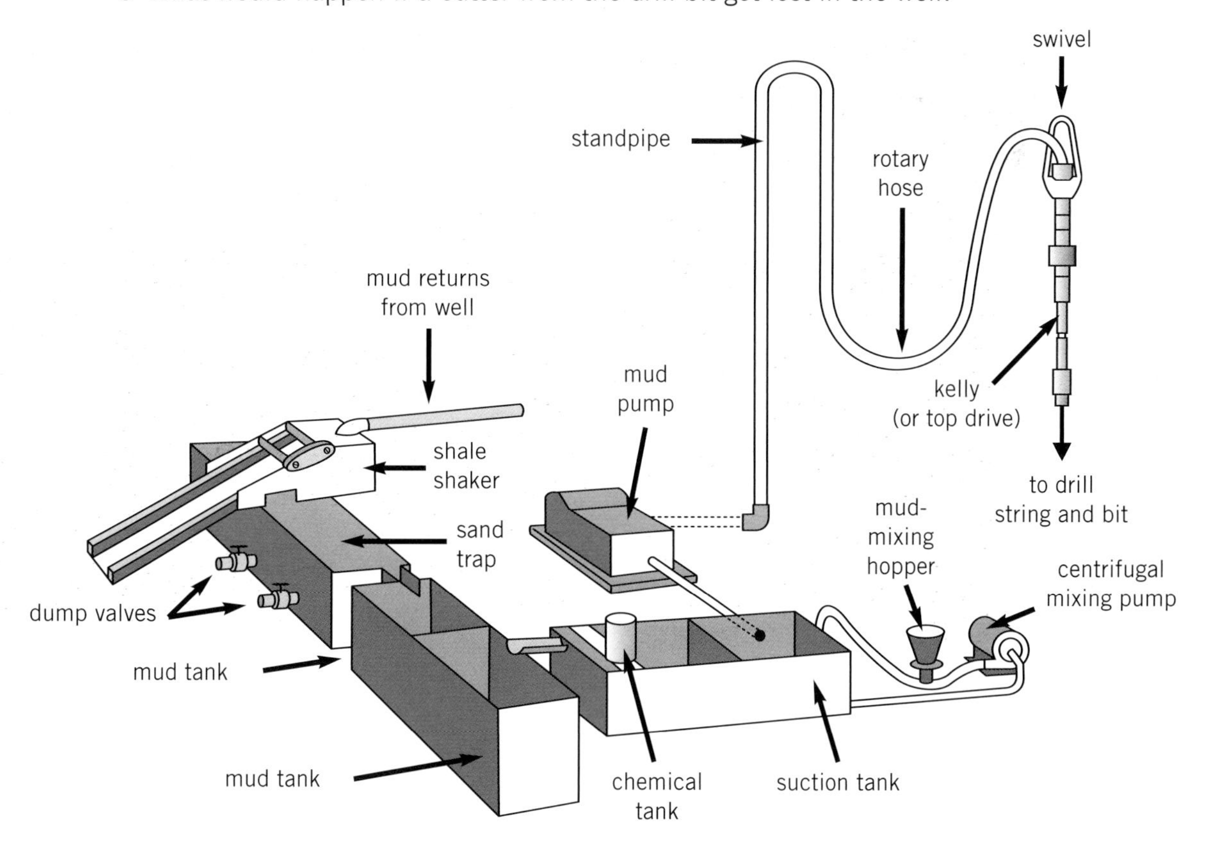

Lesson 10: Describing filters and strainers

A Write the opposites of the adjectives below.

1 solid ______________

2 fine ______________

3 offline ______________

4 worn ______________

5 increased ______________

B Choose the correct modal verbs to complete the paragraphs.

One problem for pipelines is solids in liquids. These solid particles *should / might / must* be carried in the liquid or *need to / might not / could* be formed within the pipeline and be the result of corrosion of the pipe wall. As solid particles build up, they lower the flow rate of the pipe. Solid particles *must / can / should* also cause excessive wear or equipment malfunction, so they *have to / could / may* be removed to protect equipment and extend the working life of parts.

Because of this, filters and strainers *must / don't have to / should* be fitted within pipelines. Filters are finer than strainers, so they *should / mustn't / may* be used where smaller particles need to be caught. Strainers usually form a metal screen and *must / can / could* catch larger coarse particles.

C Answer the questions below about filters and strainers.

1 Where do solid particles in pipelines come from?

____

____

____

2 Why do they need to be removed?

____

____

____

3 What is the difference between filters and strainers?

____

____

____

D Compare the filters in the diagrams below.

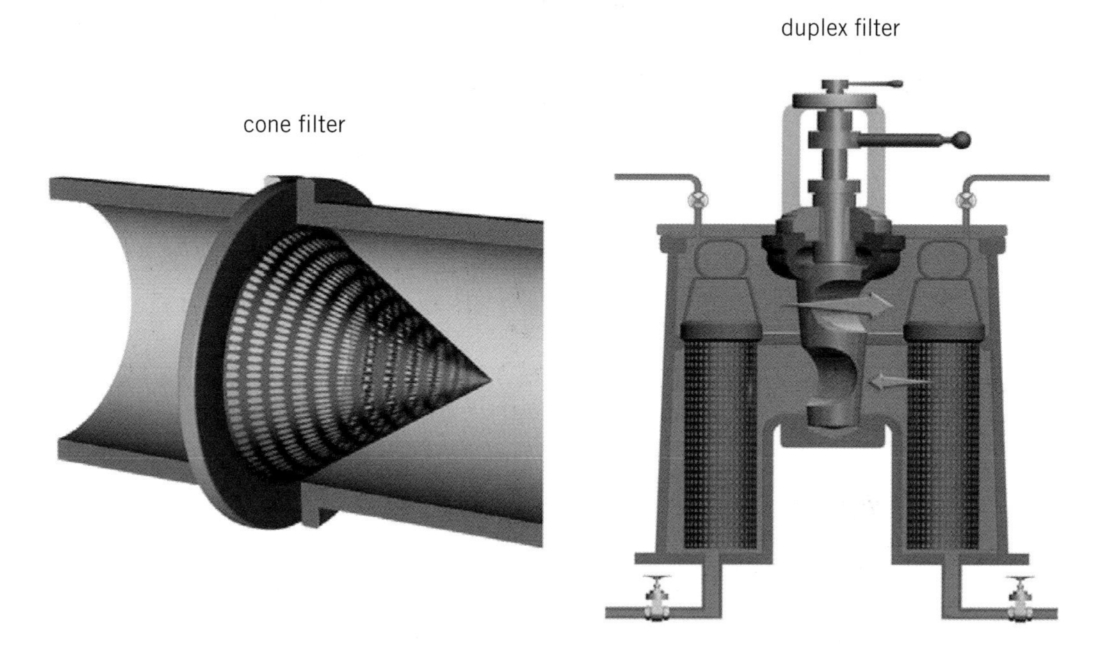

E Listen and check your answers. Complete the table with the advantages and disadvantages of each type of filter.

	Cone filters	Duplex filters
Advantages		
Disadvantages		

UNIT 9

Review: Giving advice

A Complete each sentence with a suitable ending.

1 Roustabouts *have to / don't have to* ______________________________.

2 When using filters, you *should / have to* ______________________________.

3 A mud man has to ______________________________.

4 If there is a blowout, you *should / shouldn't* ______________________________.

5 A round trip has to be made when ______________________________.

6 If you are involved in drilling, you *should / shouldn't* ______________________________.

B Work with a partner. Plan a toolbox talk on one of the topics below. Use at least one sentence from exercise A. Give your talk to another pair who have chosen a different topic.

- Working on the floor of the rig
- Mud systems
- Strainers and filters
- Drilling for oil
- Hazards on the rig

C Work in small groups. Discuss the following questions and make notes.

What would happen if ...	
an expensive item got lost in an oil well?	
someone was driving a vehicle dangerously on site?	
a new filter wasn't working properly?	
there was a fire on the rig?	
What would you do if ...	
you were always late for work?	
you needed some money quickly?	
you thought that someone was working offshore without proper training?	

UNIT 9

Assess your skills: Giving advice

Complete (✓) the tables to assess your skills.

I can ...	Difficult	Okay	Easy
• discuss rules and obligations connected with transportation and the oil industry.			
• give advice.			
• make deductions and discuss possibilities in order to troubleshoot problems.			
• understand and give toolbox talks about equipment and hazards.			
• understand technical expressions and identify formal and informal registers.			
• discuss hypothetical situations.			
• understand texts about filters, drilling and mud systems.			

I understand ...	Difficult	Okay	Easy
• the unit grammar: modal verbs for - obligation - advice - possibility			
first conditional			
second conditional			
• the unit vocabulary (see the glossary)			

If there is anything you are not sure of, ask your trainer to revise the material.

Review
Unit 1: Giving basic information

A Complete the text with the correct form of the verbs in the box.

want	like	have	work (x2)	spend

Susan [1] ________________ a degree in electrical engineering, but she [2] ________________ as an electrical engineer. She [3] ________________ as a maintenance technician, but she [4] ________________ her job and [5] ________________ to get a job in her speciality. She [6] ________________ a lot of time studying electrical engineering manuals.

B Use the verbs above to write six sentences about yourself or your colleagues.

1 ____

____

2 ____

____

3 ____

____

4 ____

____

5 ____

____

6 ____

____

C **Work with a partner.**

Student 1: Read out the time on each clock. Write the letters you hear under the corresponding clock. What are the words you have formed?

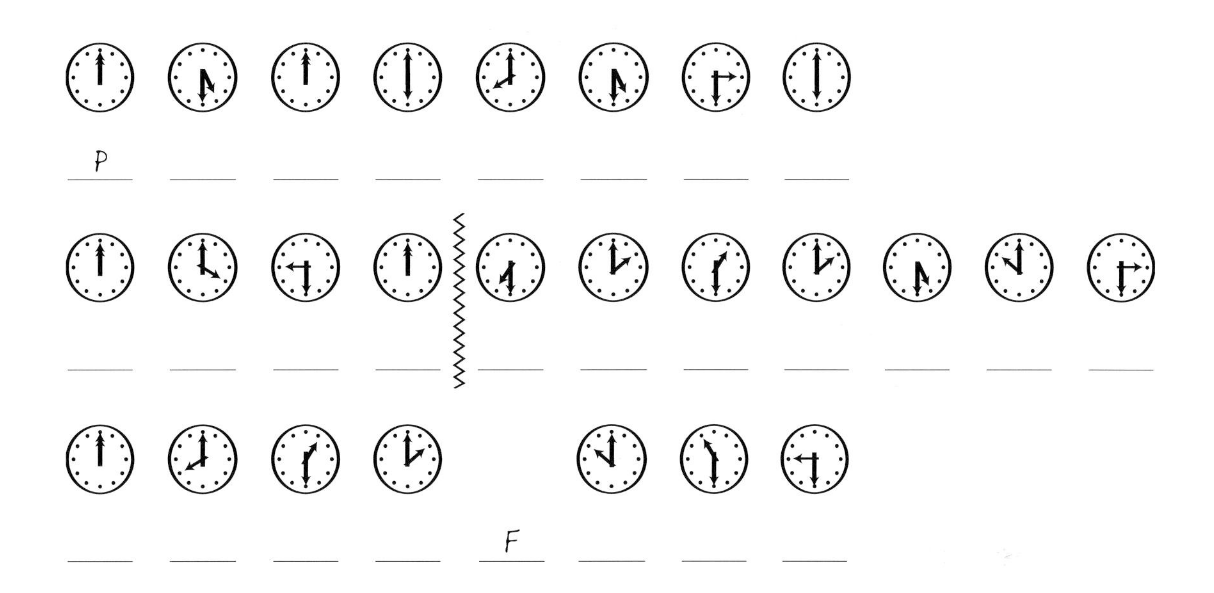

Student 2: Look at the key on page 233. Inform Student 1 of the letter that each clock represents.

Swap roles.

Student 2: Read out the time on each clock. Write the letters you hear under the corresponding clock. What are the words you have formed?

Student 1: Look at the key on page 223. Inform Student 2 of the letter that each clock represents.

Unit 2: Calculating and measuring

A Can you remember what these abbreviations refer to? Write their full form.

1 cm ______________________

2 psi ______________________

3 mph ______________________

4 kPa ______________________

5 kw ______________________

6 rpm ______________________

7 mg ______________________

8 Hz ______________________

9 lb ______________________

10 hr ______________________

B Describe the dimensions of the pictures below.

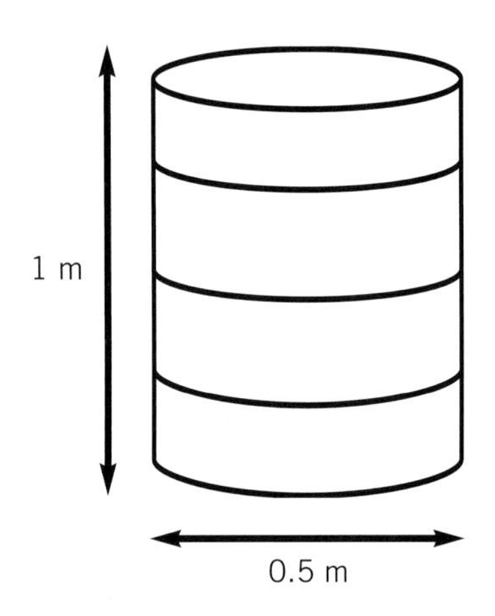

____

____

____

____

____

____

____

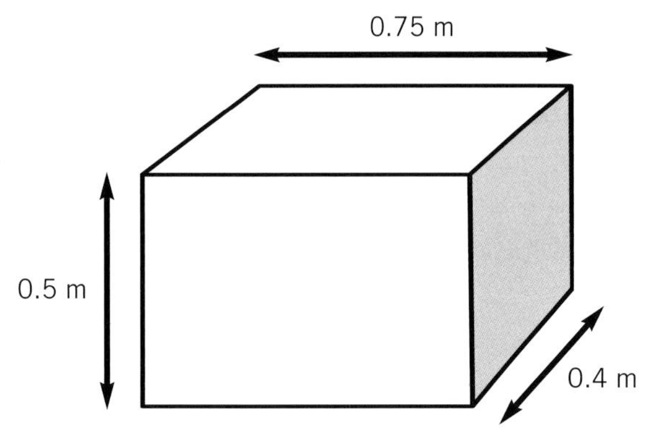

____

____

____

____

____

____

____

C Work with a partner.
Student 1: Look at the periodic table on page 223.
Student 2: Look at the periodic table on page 233.

Ask your partner for the missing information to complete the table.

What is the atomic weight of …?

What element has the atomic number …?

What is the symbol for …?

What does … stand for?

Unit 3: Describing equipment

A **Look at the information in the table below. Complete the sentences with *can* or *can't*.**

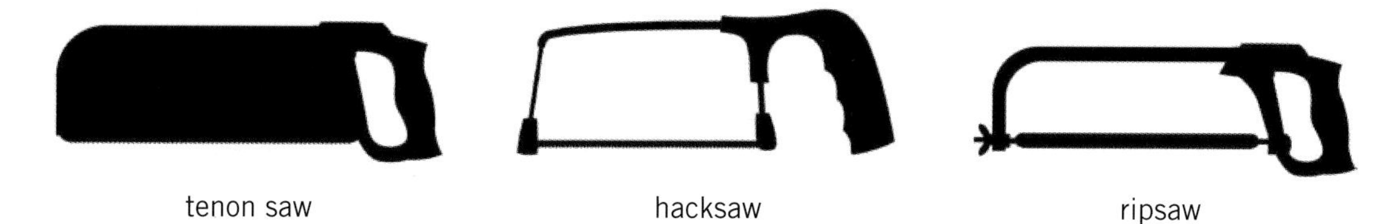

Type of saw	Metal	Wood	Thickness
tenon saw	✗	✓	2 mm–30 mm
hacksaw	✓	✗	1 mm–50 mm
ripsaw	✗	✓	30 mm–300 mm

1 A tenon saw can't cut metal 10 mm thick.

2 A hacksaw ________________ cut metal 6 cm thick.

3 A ripsaw ________________ cut wood 25 cm thick.

4 A ripsaw ________________ cut metal 15 mm thick.

5 A tenon saw ________________ cut wood 2.5 cm thick.

6 A hacksaw ________________ cut wood 5 mm thick.

B **Complete the sentences with the active or passive voice, with *can/can't* and the verbs in the box.**

use (x2)	loosen	hold	sharpen	make

1 You ________________ a drill to make a hole.

2 A drill ________________ to make a hole in metal.

3 You ________________ work securely with a vice.

4 You ________________ a bolt with a hammer.

5 A chisel ________________ with a grinder.

6 You ________________ a hole with a saw.

C Complete the text with the verbs in the correct form, active or passive.

The diagrams below [1]____________ (*show*) a double-acting, piston-type, reciprocating pump. The pump [2]____________ (*have*) two discharge valves, D.A. and D.B., and two suction valves, S.A. and S.B. When the piston [3]____________ (*move*) from left to right, as shown in diagram A, the liquid [4]____________ (*be*) drawn in through the suction valve S.A. At the same time, liquid [5]____________ (*force*) out through the discharge valve D.B. When the piston [6]____________ (*reverse*) and [7]____________ (*move*) from right to left, as in diagram B, liquid [8]____________ (*draw*) in through the suction valve S.B. At the same time, liquid [9]____________ (*force*) out through the discharge valve D.A.

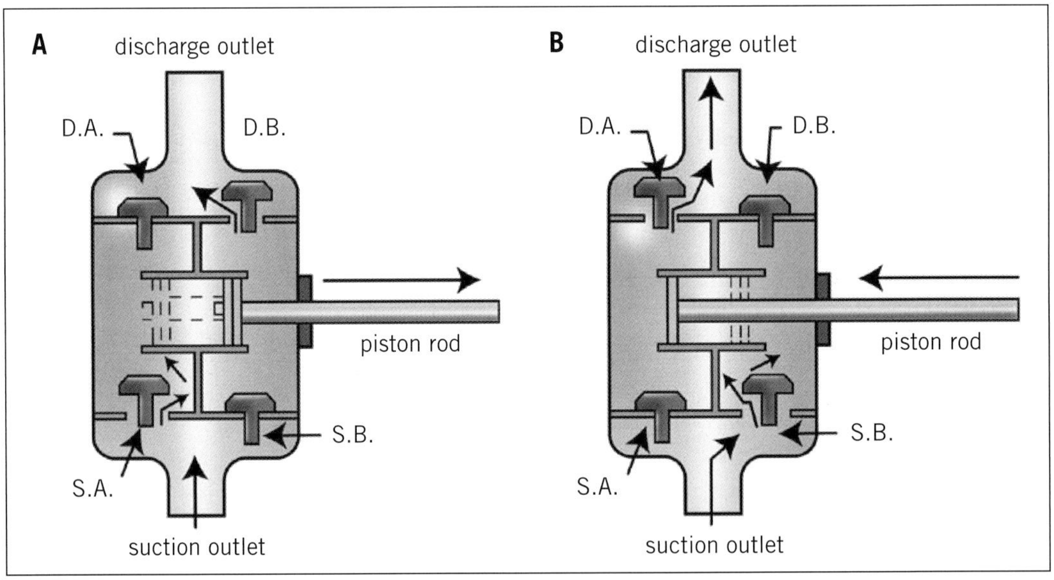

With this arrangement, both sides of the piston [10]____________ (*take*) part in the pumping action, and liquid [11]____________ (*discharged*) when the piston [12]____________ (*move*) in either direction. Therefore, this type of pump [13]____________ (*call*) double-acting.

D Decide whether the sentences below are true (T) or false (F).

1 When the piston moves to the right, valve S.A. is open. ☐

2 Valves S.B. and D.B. can be open at the same time. ☐

3 Liquid moves through the pump on every stroke. ☐

4 Liquid is drawn in through the top of the pump. ☐

Unit 4: Giving instructions and warnings

A Look at the picture of a cassette player. Identify the following items.

the play button	the pause button	the rewind button
the fast-forward button	the stop/eject button	the volume dial

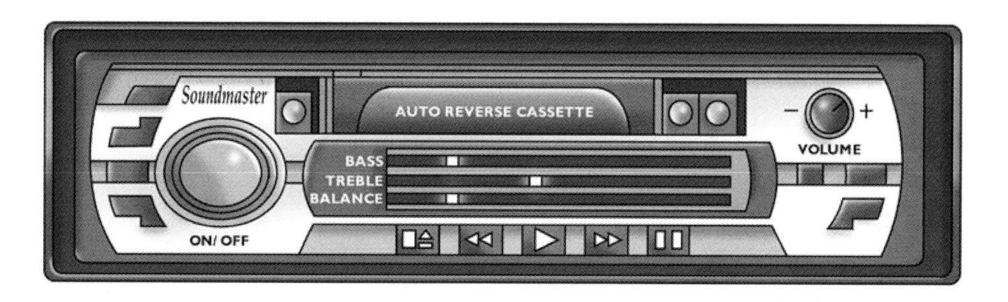

B Work with a partner. Discuss how the machine works. Then complete the instructions below.

1 Press the ___________ button to turn the cassette player on.

2 _______ the _______ button again to turn it off.

3 _______ the play _______ to play the cassette.

4 _______ the stop button to _______ the cassette.

5 ______ it again to eject the cassette.

6 ______ the _______ dial clockwise to increase the volume.

7 _______ the _______ _______ - _______ to decrease the volume.

8 Move the bass switch to the left to _______ _______ _______.

9 Move the treble _______ to the right to _______ _______ _______.

10 _______ the pause _______ _______ _______ _______ _______.

C Work with a partner. Write a set of instructions for using a tool or piece of electrical equipment of your choice.

D **Think of some warnings for using electrical equipment. Make sentences using the words in the box.**

always	never	don't	be careful to	you must	you can't
remove	check	avoid	turn off	disconnect	plug in

Always turn off electrical equipment at the plug.

____

____

____

E **Look at the picture. What does this person do? What is he wearing? What is he doing? Write a paragraph describing him.**

____

____

____

____

____

Unit 5: Describing systems

A Complete the definitions with the correct words.

1 A gas turbine is an internal combustion engine *who* / *which* consists of a compressor, a combustion chamber and a turbine wheel.

2 A ____________________ is an injury, such as a bruise, *which* / *where* the skin is not broken.

3 A ____________________ or ____________________ is someone *who* / *which* helps a crane operator by giving him or her directions.

4 ____________________ are orange and white objects *who* / *that* are placed around an area to stop people going there.

5 ____________________ are gloves *which* / *who* protect the hands against hazardous chemicals.

6 ____________________ are diagrams *where* / *which* show complex industrial systems in a simple way.

B Work with a partner.
Student A: Look at page 224.
Student B: Look at page 234.

Make up definitions for the words in your list using relative clauses. Read your definitions to your partner and see if they can guess the words.

C Ask your partner how often they use/see/go to the objects/people/places from exercise B.

How often do you wear safety goggles?

D Look at the diagram of a gas turbine. Discuss what you think happens in each section.

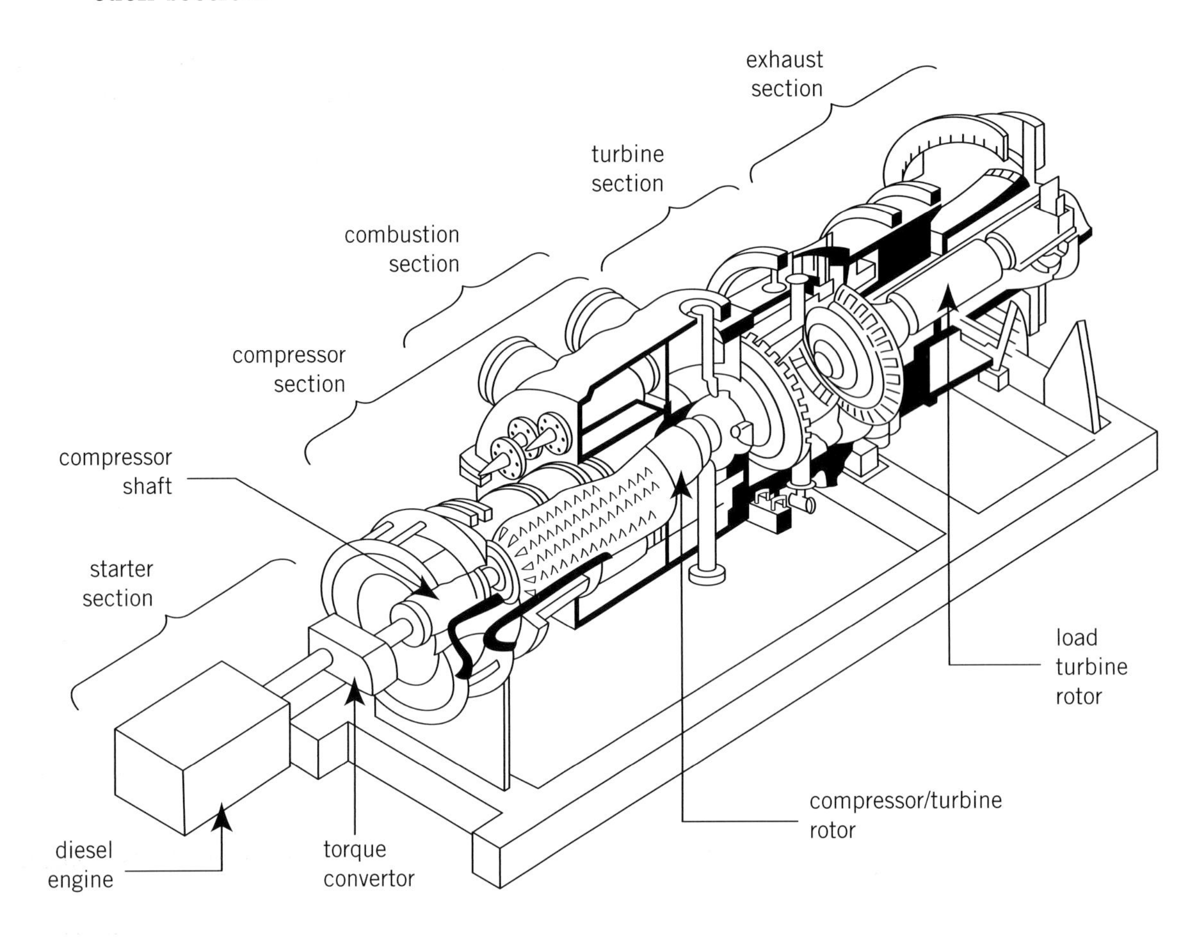

E Work with a partner.
Student A: Look at page 224.
Student B: Look at page 234.

Read the description of the gas turbine. Ask your partner questions to find which words complete the gaps in your text.

Unit 6: Talking about safety

A Each conditional sentence below contains a mistake. Underline them and rewrite them correctly. Which sentences use the zero conditional?

1 If you will press the emergency stop button, the system shuts down.

____

2 If it rains, we will don't dig the trench.

____

3 If rubber is cooled to –200°C, it become brittle and breaks.

____

4 If you not wear safety gloves, you might injure your hand.

____

5 The platform will be installed in time if the sea will be calm.

____

B Complete each sentence in a logical way with the zero or first conditional.

1 If I am late for work, ______________________________.

2 If I work rotation, ______________________________.

3 I will get a promotion if ______________________________.

4 If there is a problem, ______________________________.

5 If you don't speak good English, ______________________________.

6 If you don't wear the correct PPE, ______________________________.

C Complete the paragraph with the verbs in the correct past tense form.

Yesterday afternoon, I [1] ______________________ (*work*) on site. I [2] ______________________ (*change*) a faulty valve because there [3] ______________________ (*be*) a problem with the flow. While I [4] ______________________ (*remove*) the faulty valve, my pipe wrench [5] ______________________ (*slip*) and I [6] ______________________ (*cut*) my left hand. The wound 7 ________________ (*bleed*) heavily, so I

8 ________________ (*wrap*) it in a handkerchief and

9 ________________ (*apply*) pressure. A colleague

10 ________________ (*take*) me to the medical block and a doctor

11 ________________ (*examine*) me. He 12 ________________ (*tell*)

me I 13 ________________ (*need*) stitches. While he

14 ________________ (*stitch*) the wound, I 15 ________________

(*feel*) faint, but when he 16 ________________ (*finish*),

I 17 ________________ (*feel*) better.

D Test your partner on the past form of the following verbs.

1 work ________________

2 do ________________

3 see ________________

4 get ________________

5 take ________________

6 go ________________

7 have ________________

8 meet ________________

E Write ten questions for a partner with the verbs in the past tense from exercise D and the question words in the box below.

Why?	When?	Where?	What?	Who?
How many?	How often?	How long?		

When did you start working in the oil industry?

What were you doing at 10 o'clock last night?

Unit 7: Making comparisons

A **Read the sentences below about different metals. Decide whether they are true (T) or false (F). Rewrite the false ones so that they are true.**

1 Gold is heavier than steel. ☐
2 Zinc has a higher tensile strength than copper. ☐
3 Silver is more lustrous than iron. ☐
4 Copper is more corrosive than iron. ☐
5 Lead is more ductile than tungsten. ☐
6 Steel is more brittle than mercury. ☐
7 Titanium is harder than lead. ☐
8 Silver has higher thermal conductivity than copper. ☐

B **Compare the two vehicles in the diagrams. What are the main differences between them? Make notes under the headings below.**

Size and shape	
Age/Condition	
Safety	
Uses/Efficiency	
Other	

C **Choose one of the groups below. Write notes comparing the three things. Use some of the headings from exercise B.**

1 mercury thermometers / pentane thermometers / alcohol thermometers
2 diesel vehicles / petrol vehicles / gas-powered vehicles
3 foam pigs / metal pigs / plastic pigs
4 transporting oil by train / transporting oil by lorry / transporting oil by boat

D **Give a short talk to a partner or to the class about the three items you have compared.**

Unit 8: Describing processes and procedures

A Complete the conversation with the verbs in the past simple or present perfect tense.

1 **A:** __________ you __________ (*have*) previous experience in the oil industry?

2 **B:** Yes. I ____________________ (*work*) onshore and offshore as an operator.

3 **A:** When __________ you __________ (*work*) offshore?

4 **B:** From 1999 to 2002. In 2002, I____________________ (*work*) onshore.

5 **A:** Why __________ you __________ (*change*) from offshore to onshore?

6 **B:** A better position ____________________ (*become*) available onshore.

7 **A:** And __________ you __________ (*like*) working offshore?

8 **B:** I ____________________ (*like*) the people, but not the rotation. I sometimes ____________________ (*find*) it difficult to work six weeks.

9 **A:** And __________ you __________ (*make*) any plans to change your position again?

10 **B:** Not at the moment.

B Look at the pictures and discuss what has happened.

1

2

3

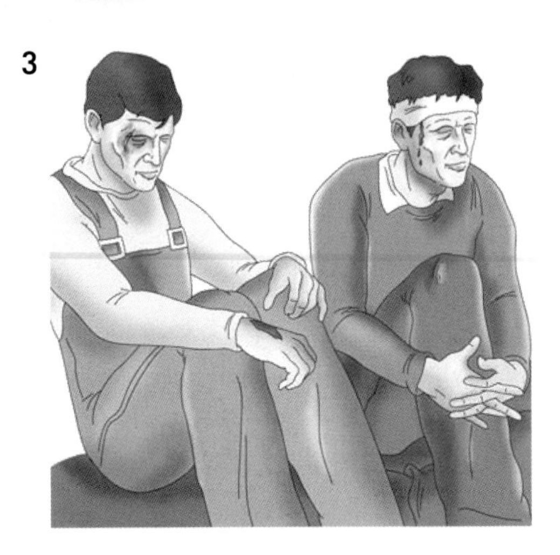

4

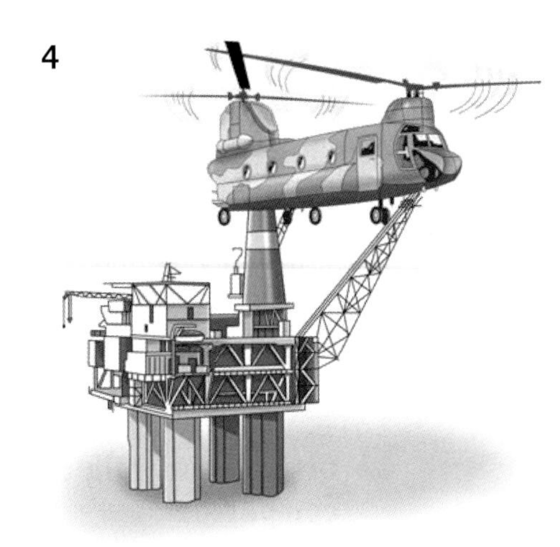

C **Put the sentences below into a logical order to make a newspaper report about the incident.**

Accident at Britain's largest offshore gas storage facility

1 22 essential crew members have stayed on the rig to maintain the facility. ☐

2 The facility has now been shut down and made operatively safe. ☐

3 Thick white smoke was seen coming from a gas storage facility off the East Yorkshire coast this morning. ☐

4 At lunchtime, a spokesperson commented, "A small fire has been extinguished." ☐

5 Two workmen were reported to be suffering from minor burns and shock. They were immediately taken to hospital. ☐

6 60 non-essential crew members were airlifted to shore this afternoon. ☐

D **What else would you like to know about the incident? Write four questions to ask your instructor. They will try to answer them.**

____

____

____

____

____

____

____

____

Unit 9: Giving advice

A Work in small groups. Discuss your opinions about the statements below.

1 In the future, it is likely that oil companies will use systems and chemicals that are less damaging to the environment.

2 It would be very useful if we knew when oil reserves were going to finish.

3 The incidence of fires and explosions on oil rigs is likely to increase unless there is more safety training.

4 If an oil company offered me a lot of money to work in Antarctica, I would go.

B Read the toolbox talk and complete the gaps with suitable modal verbs.

I want to talk to you about an area of major environmental concern: ground contamination. Ground contamination [1]____________ have a very long-term impact. Even small spills [2]____________ have a major impact on the local area and accumulate to become a serious problem that leads to health problems.

Ground contamination results from poor storage, handling and disposal of chemicals and fuel. It [3]____________ be avoided by using the correct control techniques.

First, you [4]____________ maintain good general practice – all waste [5]____________ be put into the right containers. You [6]____________ clean up any spills immediately and you [7]____________ report them as soon as you can.

Also, containers and barrels containing petroleum, oil or chemicals [8]____________ always be returned to the correct area at the end of the day, and [9]____________ be covered. Storage tanks [10]____________ have adequate binding around them and be covered to avoid potential spills.

Empty drums and containers [11]____________ be disposed of in the correct waste facilities, as there is always some residue chemical that [12]____________ contaminate the ground.

If you keep to these guidelines and take care when using chemicals, spills [13]____________ be reduced and the threat of contamination minimized. Remember, you are responsible not only for yourself, but also for your environment and the effects of your actions on other people.

C Look back through the book. Choose a topic for a toolbox talk. Structure your talk in the following way.

1 Write a list of important vocabulary and expressions connected with the topic.

2 Write three or four headings summarizing the main points you want to make in your talk.

3 Make notes under each heading. Try to include most of the words from point 1.

4 Practise giving your talk to a partner.

5 Give your talk to the rest of the group.

Pairwork activities
Student 1

Unit 2, Lesson 1, Exercise D
Page 25

Take it in turns to say the numbers. Write down the numbers that your partner says. Student 1 starts first.

1 12
2 2,143
3 321,654
4 87,645,231
5 1,243,657,890
6 2,060

Unit 2, Lesson 4, Exercise F
Page 31

Take it in turns to read your sentences aloud. See if your partner can identify the three sentences which are incorrect.

1 Would you like some cakes? (*Correct. 'Cakes' refers to a number of small cakes.*)
2 Long hairs should be tied back when you are working. (*Incorrect. When we talk about the hair on someone's head, we use the singular.*)
3 How much oil does Kuwait produce? (*Correct. 'Oil' is uncountable.*)
4 Be careful! There's a glass on the floor. (*Correct. 'A glass' refers to a drinking glass.*)
5 There isn't much people on the rig. (*Incorrect. 'People' is plural.*)
6 We had few problems, but we've sorted them out now. (*Incorrect. It should be 'a few problems'.*)
7 It's good to have plenty of exercise. (*Correct. 'Exercise' can be uncountable if we refer to it in general terms.*)

Unit 2, Lesson 9, Exercise A
Page 40

Check your answers with a partner.

1 The freezing point of water is 0°C.
3 The normal temperature of the human body is 37°C.
5 0°K is –273°C.

Unit 2, Lesson 9, Exercise E
Page 41

Swap information so that you can complete the right-hand column of the table.

2 Prague–Mumbai: 9,334 miles
4 Bangkok–Tokyo: 2,849 miles
6 Cairo–Singapore: 5,127 miles
8 Paris–Hong Kong: 8,193 miles

Unit 3, Lesson 3, Exercise D
Page 50

Describe the position of the tools. Find eight differences between the pictures without looking at your partner's picture.

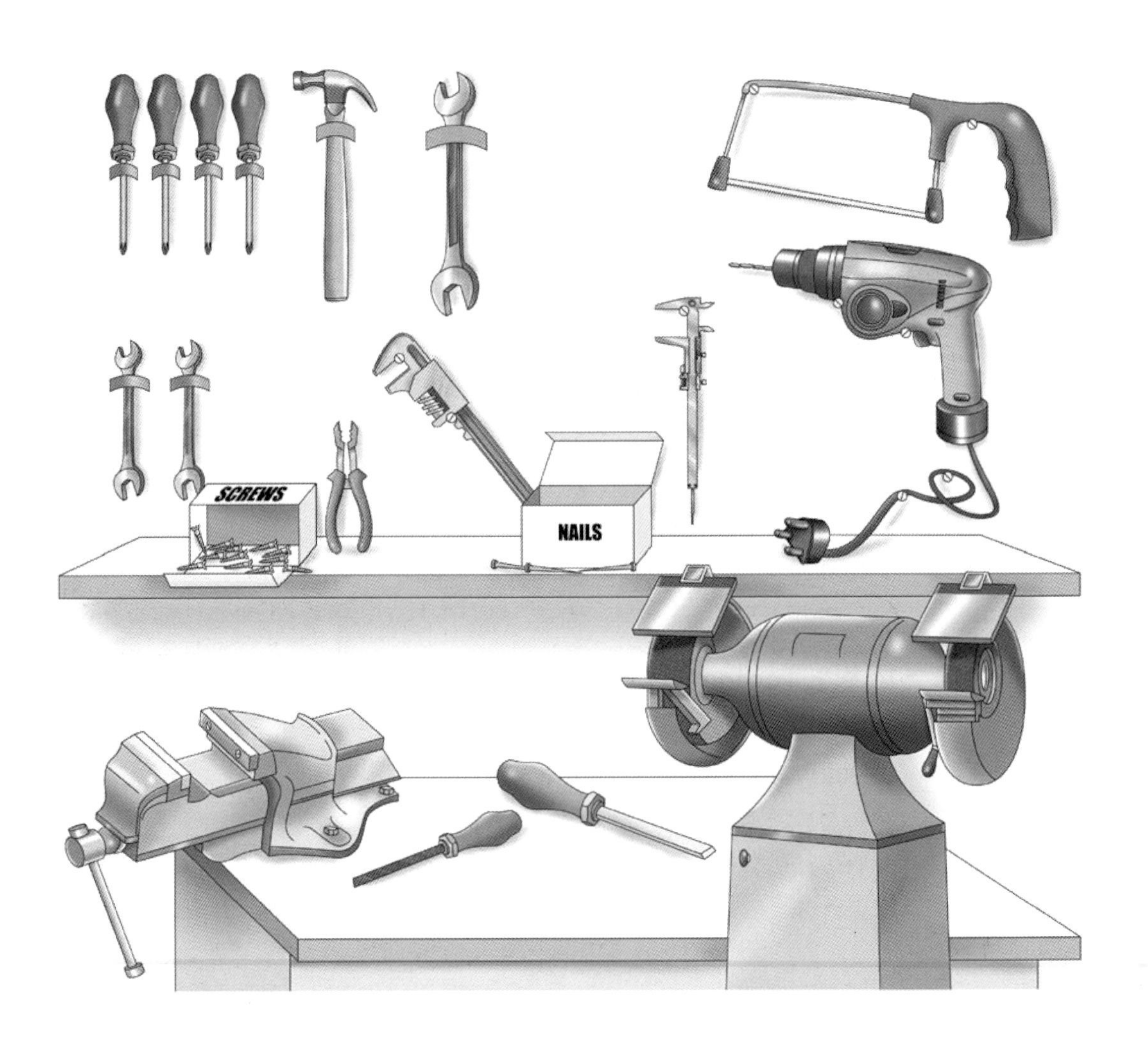

Unit 3, Review, Exercise A
Page 66

Take it in turns to tell your partner about your picture. Find six differences between the pictures without looking at your partner's picture.

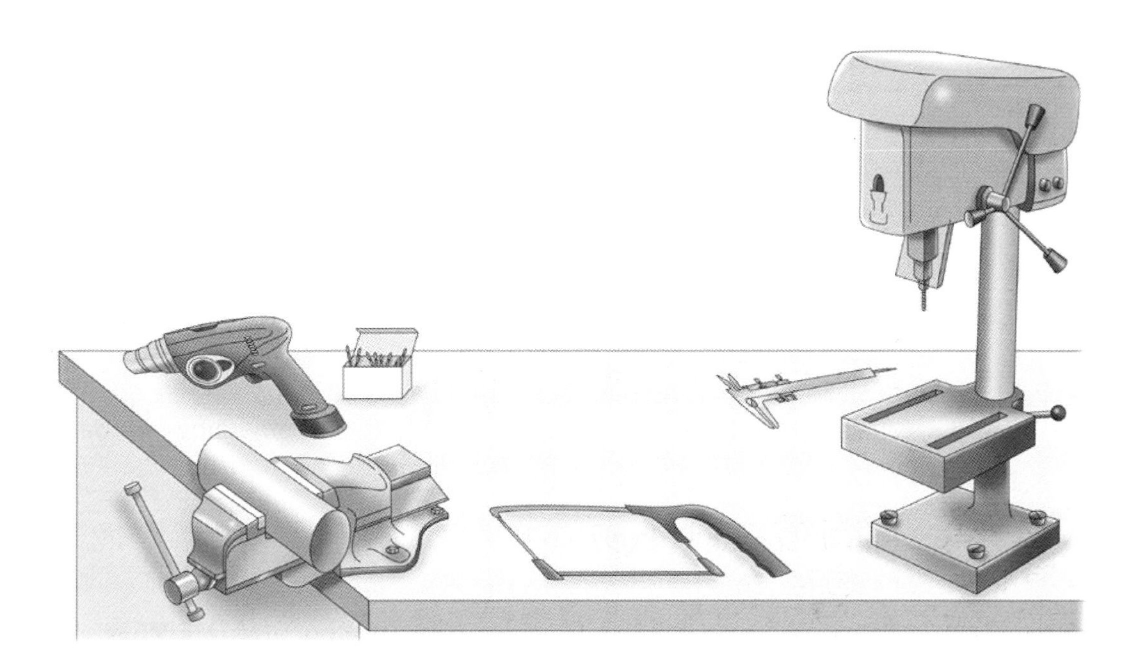

Unit 3, Review, Exercise B
Page 66

Answer the questions for the valve you read about.

Valve A – Butterfly valve
Butterfly valves are shut-off valves. This means they can stop the flow of gas or oil completely. They are used in low-pressure pipelines (gas or oil). The valve consists of a circular flap or disc the same diameter as the internal diameter of the pipeline. The disc can pivot or turn through 90° in the pipeline. When the disc is parallel to the flow, it is fully open and the flow is uninterrupted. When the disc is at a right angle to the flow, it is fully closed. It is operated by a handle that is turned to set the valve anywhere between fully open and fully closed.

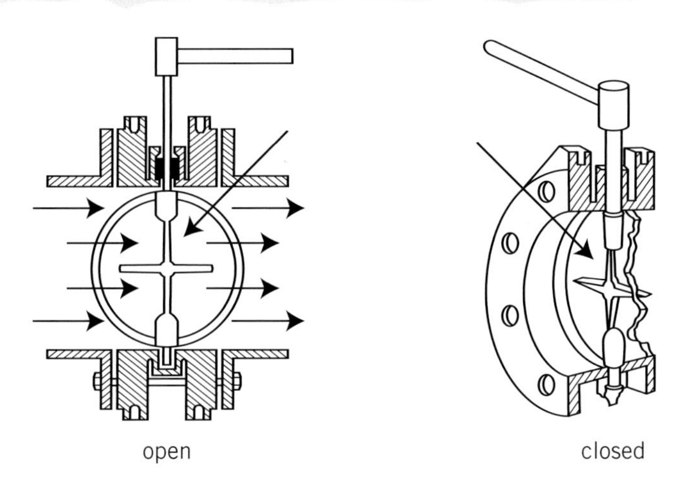

Unit 4, Lesson 3, Exercise C
Page 73

Check that your partner's instructions are correct.

1 Changing a fuse in a plug
Check that the power is off.
Take out the old fuse and dispose of it carefully.
Check that the new fuse is the correct voltage.
Push it into the fuse holder.
Turn on the power and check that the plug works.

2 Climbing up a ladder
Make sure you are wearing appropriate shoes or boots.
Position the ladder securely. (Ask someone else to hold it, if possible.)
Test each step as you climb.
Make sure your feet are in the middle of the steps.
Move the ladder every time you have to reach something far away.

Unit 4, Lesson 5, Exercise E
Page 77

Tell your partner what is happening in your pictures. They should respond with an appropriate warning.

1

2

Unit 4, Review, Exercise C
Page 88

Take it in turns to describe the picture without letting your partner see it. Your partner should say what they think is happening in your picture.

Unit 5, Lesson 2, Exercise D
Page 93

Read the bullet points below. Try to think of disadvantages of open-loop systems and advantages of closed-loop systems. Compare your ideas with a partner.

	Advantages	Disadvantages
Open-loop systems	• Cheap. • Easy to install. • Easy to operate. • Easy to maintain.	• ________ • ________ • ________
Closed-loop systems	• ________ ________ • ________ ________ • ________	• More complex to repair. • Danger of accidents if the controller does not register a problem. • If the instrument fails, flow operates outside preset levels.

Unit 5, Lesson 7, Exercise B
Page 103

Look at the key and complete table 1 on page 103 with abbreviations and full words. Ask your partner to help you complete table 2.

tank	T	compressor	K
pump	P	filter/strainer	S
heat exchanger	E	alarm	A
controller	C	drain	D

Unit 5, Review, Exercise C
Page 110

Look at the diagram. Ask your partner these questions.

1 What are the main components of the system?
2 What is the purpose of the sphere-deflector assembly?
3 What does the meter measure?

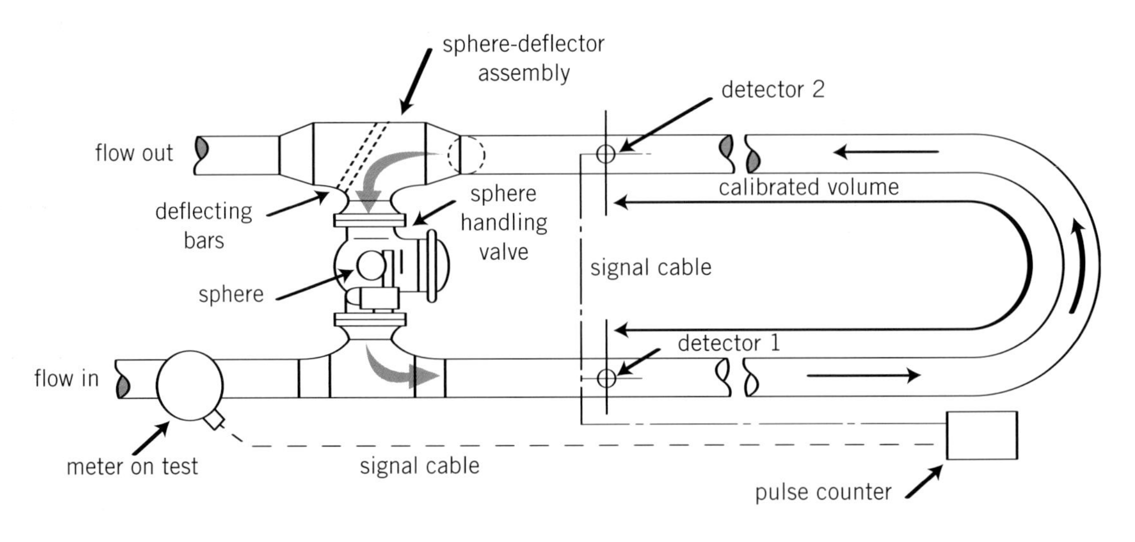

Unit 6, Lesson 6, Exercise B
Page 122

Complete the past tense table. Check your answers with your partner.

come		adjust	adjusted
control		be	was/were
hear		call	called
initiate		increase	increased
reach		shut down	shut down
try		stop	stopped

Unit 6, Review, Exercise B
Page 132

Read the incident report.

Two members of the seismic crew were in an accident at 4.00 p.m. on 22nd September, 2005. An all-terrain vehicle (ATV-quad) driver lost control of his machine as it moved down a steep hill at a high rate of speed. The ATV crashed into a tree and the driver fractured his skull. Fortunately, the passenger riding on the vehicle only suffered minor cuts and abrasions. The driver and passenger were riding on an ATV vehicle that was designed to carry only one person. Neither the driver nor the passenger were wearing safety headgear or eye protection. In addition, the vehicle's brakes were not functional as the braking system was low on fluid. The passenger was able to phone for help and the driver was taken to hospital.

Unit 7, Review, Exercise D
Page 154

Read about different types of lubrication. Swap information about the advantages of each type, then discuss and write down some disadvantages.

Lubrication is vital – it reduces the friction between moving parts so they can work smoothly, which reduces wear. There are different grades of lubricant for different purposes.

Lubricant	Advantages	Disadvantages
Grease	• ______ • ______ • ______ • ______	• Difficult to check amount. • ______ • ______ • ______
Oil	• Can be cooled. • Can be replaced easily. • Can be recirculated. • Can pass through small openings.	• ______ • ______ • ______ • ______

Unit 8, Lesson 10, Exercise A
Page 174

Read the description of how a centrifuge works and label as much of the diagram below as you can. Then swap information with your partner and finish labelling the diagram.

This diagram shows a hydrocyclone centrifuge. The liquid enters the *produced water inlet* at the top of the mechanism. It passes through the *inlet volute chamber*, then enters the *tapering section* of the centrifuge. The liquid is forced through a coil known as the *concentric reducer* at high speed, forming a vortex.

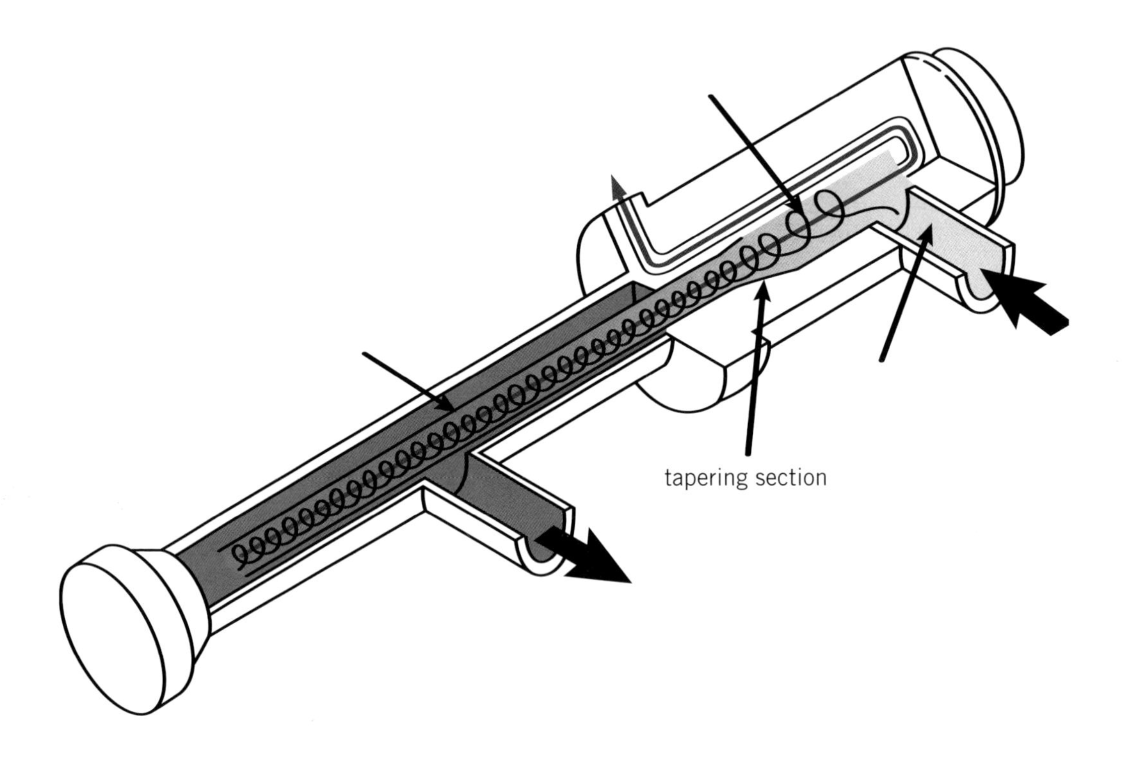

Review pairwork activities
Student 1

Review, Unit 1, Exercise C
Page 201

When your partner says a time, read out the corresponding letter.

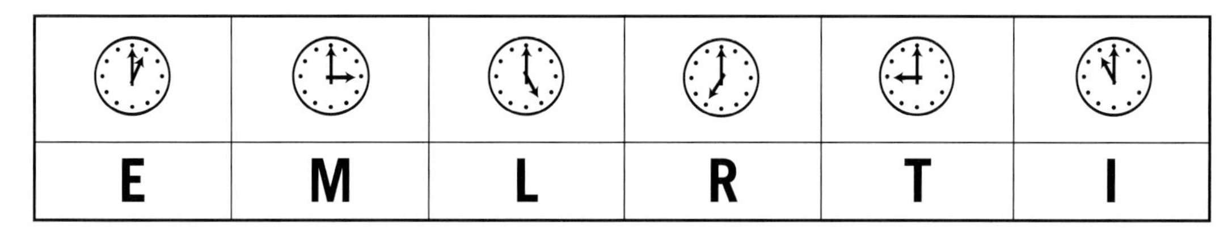

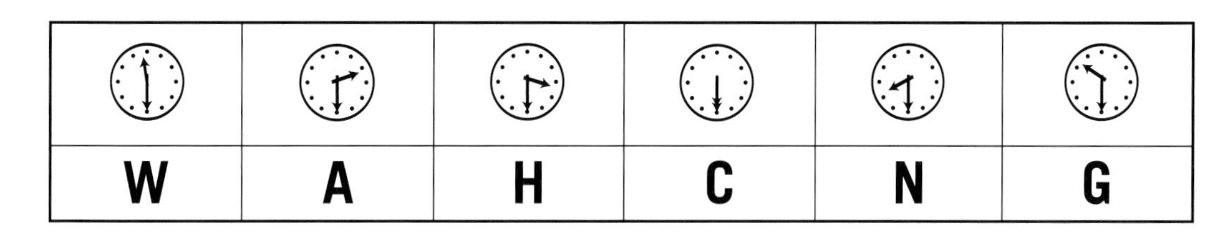

Review, Unit 2, Exercise C
Page 202

Ask your partner for the missing information to complete the table.

Atomic number	Symbol	Name	Atomic weight
1			1.00794
2	He	Helium	
3	Li	Lithium	
4			9.012182
5	B	Boron	
6	C	Carbon	
7		Nitrogen	14.00674
8	O	Oxygen	
9	F	Fluorine	
10	Ne	Neon	
11	Na	Sodium	
12	Mg	Magnesium	24.3050
13	Al	Aluminium	26.981538
14	Si	Silicon	28.0855
15	P	Phosphorus	
16	S	Sulphur	32.066
17	Cl	Chlorine	35.4527
18			39.948

Atomic number	Symbol	Name	Atomic weight
21			
22	Ti	Titanium	47.867
23			50.9415
24	Cr	Chromium	51.9961
25	Mn	Manganese	54.938049
26			
27	Co	Cobalt	58.933200
28			
29			
30			65.39

Review, Unit 5, Exercise B
Page 207

Make up definitions for the words in your list using relative clauses. Read your definitions to your partner and see if they can guess the words.

1 safety goggles
2 a foreman
3 a crane
4 an oilfield

Review, Unit 5, Exercise E
Page 208

Read the description of the gas turbine. Ask your partner questions to find which words complete the gaps in your text.

A gas turbine burns fuel to create rotational energy.

1 First, air is drawn into the compressor. It gets hot due to the compression.

2 The compressed air is fed into the ____________ in the combustion section. Here, it is mixed with fuel gas and it is ignited.

3 The flames from the ignited fuel/air mixture go through ____________ into other burner cans.

4 The burning gas becomes more pressurized and is directed onto the compressor turbine disc.

5 The ____________ rotates from the force of the gas. The compressor is on the same shaft as the turbine disc, so this rotates also. This ensures that more air is drawn into the compressor.

7 After passing through the turbine disc, the hot gas strikes the ____________. The load turbine rotates the load rotor shaft, which is connected to a pump.

8 The hot gas is no longer needed, so it leaves the engine through ____________.

Pairwork activities
Student 2

Unit 2, Lesson 1, Exercise D
Page 25

Take it in turns to say the numbers. Write down the numbers that your partner says. Student 1 starts first.

1 11
2 132
3 12,354
4 4,213,756
5 123,456,789
6 1,050

Unit 2, Lesson 4, Exercise F
Page 31

Take it in turns to read your sentences aloud. See if your partner can identify the three sentences which are incorrect.

1 Would you like some cake? (*Correct. 'Cake' refers to part of a large cake.*)
2 Long hair should be tied back when you are working. (*Correct. When we talk about the hair on someone's head, we use the singular.*)
3 How much oil wells are there in Kuwait? (*Incorrect. Oil wells are countable.*)
4 Be careful, there's some glass on the floor. (*Correct. 'Glass' refers to pieces of glass.*)
5 There isn't many people on the rig. (*Incorrect. 'People' is plural.*)
6 We had a few problems, but we've sorted them out now. (*Correct. 'A few' is used in positive sentences.*)
7 It's good to have too much exercise. (*Incorrect. 'Too much' implies more than is good.*)

Unit 2, Lesson 9, Exercise A
Page 40

Check your answers with a partner.

2 The boiling point of water is 100°C.
4 The surface of the Sun is 5,500°C.
6 32°F is 0°C.

Unit 2, Lesson 9, Exercise E
Page 41

Swap information so that you can complete the right-hand column of the table.

1 London–Moscow: 1,557 miles
3 Istanbul–Tehran: 1,270 miles
5 Kuwait–New Delhi: 1,755 miles
7 Rome–Dubai: 2,696 miles

Unit 3, Lesson 3, Exercise D
Page 50

Describe the position of the tools. Find eight differences between the pictures without looking at your partner's picture.

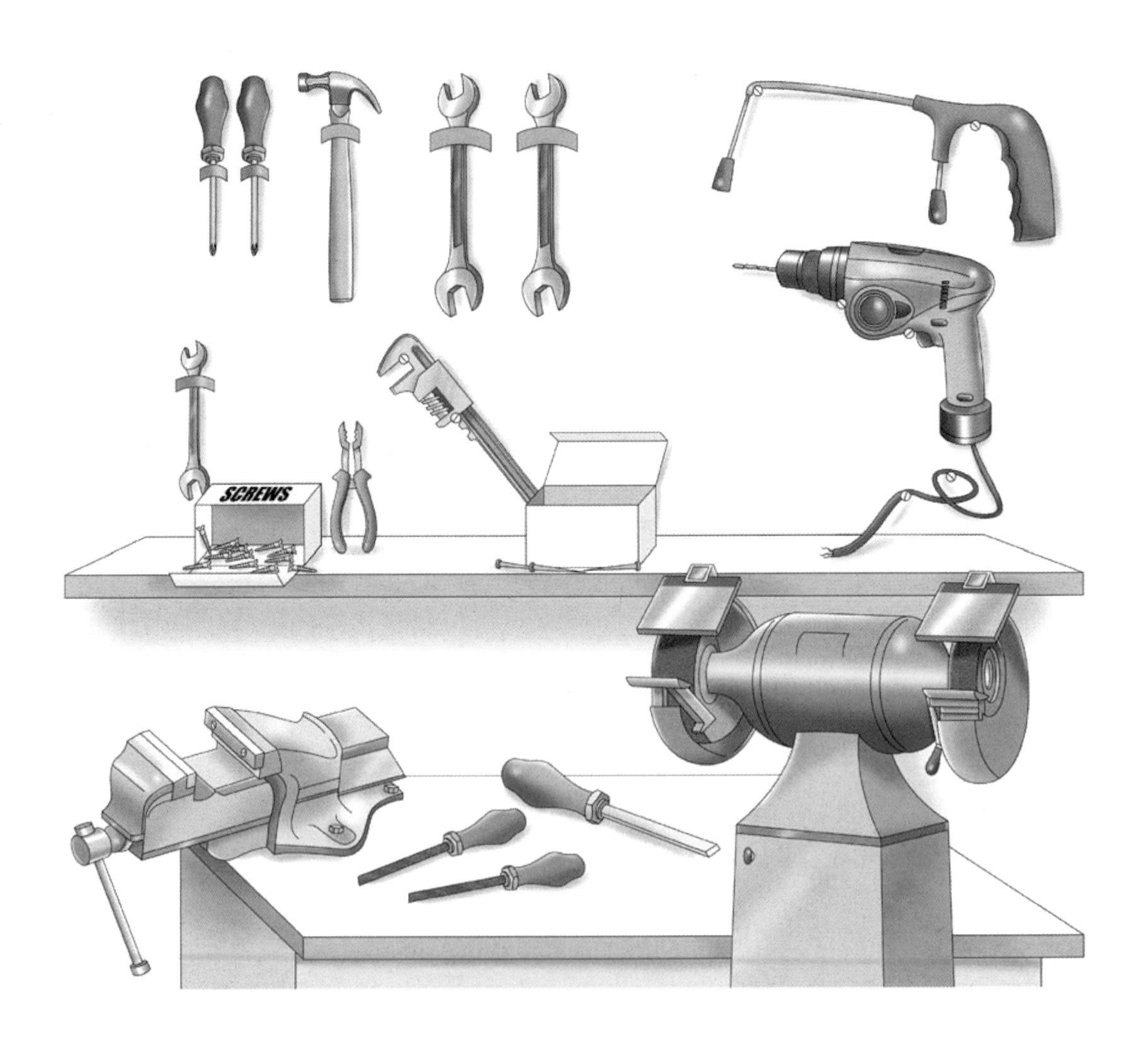

Unit 3, Review, Exercise A
Page 66

Take it in turns to tell your partner about your picture. Find six differences between the pictures without looking at your partner's picture.

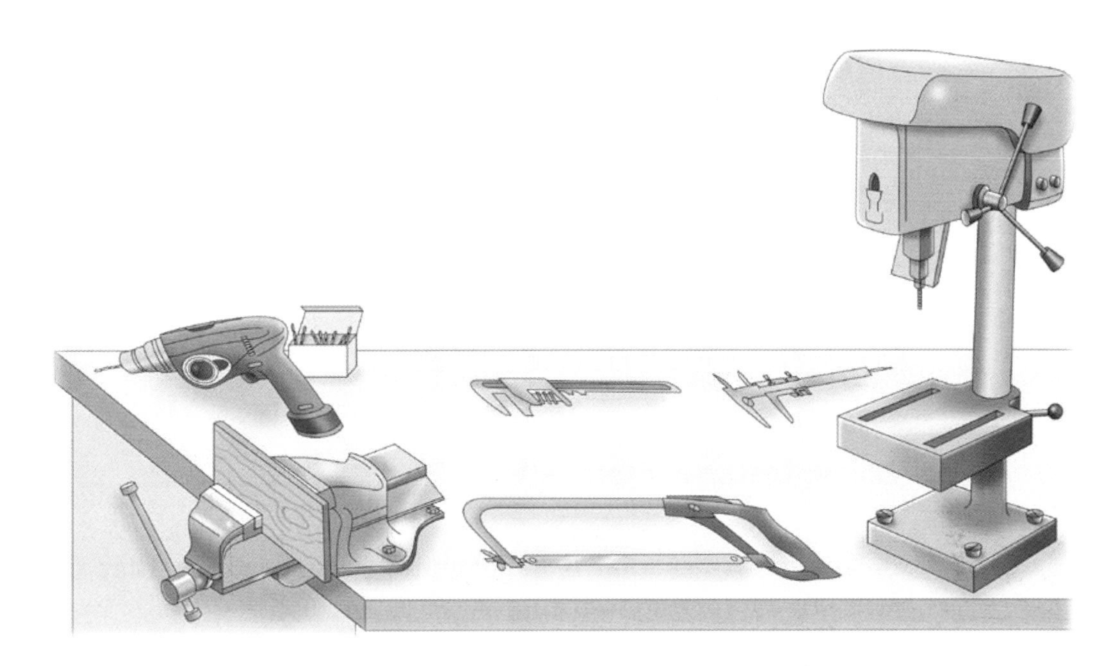

Unit 3, Review, Exercise B
Page 66

Answer the questions for the valve you read about.

Valve B – Swing check valve

A swing check valve is a check valve. This means that the valve stops liquid or gas from flowing backwards. In this type of valve, the opening element is a disc attached to a hinged arm. Pressure from the liquid upstream causes the hinged disc to rise, opening the valve. Higher downstream, pressure causes the valve to close. Swing check valves can be installed vertically or horizontally. The pressure of liquid flowing through the valve holds the disc open, allowing full flow. If flow stops, or the pressure downstream of the valve becomes higher than the pressure upstream, this makes the hinged arm swing down. The disc covers the passage and closes the valve. The higher downstream pressure against the disc keeps it tightly closed and stops liquid or gas from flowing back through the valve.

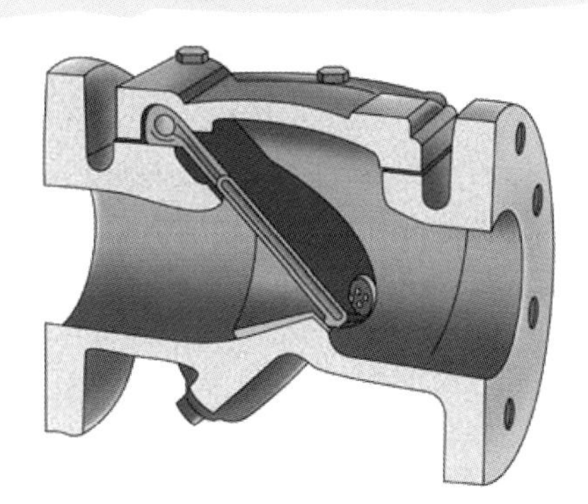

Unit 4, Lesson 3, Exercise C
Page 73

Check that your partner's instructions are correct.

1 Changing a light bulb
Check that the power is off.
Take out the old bulb and dispose of it carefully.
Screw in the new bulb.
Turn on the power and check that the light works.

2 Cutting a piece of pipe
Place the piece of pipe in a vice.
Adjust the vice to secure the pipe in place.
Use a hacksaw to cut the pipe.
Loosen the vice and remove the piece of pipe.

Unit 4, Lesson 5, Exercise E
Page 77

Tell your partner what is happening in your pictures. They should respond with an appropriate warning.

Unit 4, Review, Exercise C
Page 88

Take it in turns to describe the picture without letting your partner see it. Your partner should say what they think is happening in your picture.

Unit 5, Lesson 2, Exercise D
Page 93

Read the bullet points below. Try to think of advantages of open-loop systems and disadvantages of closed-loop systems. Compare your ideas.

	Advantages	Disadvantages
Open-loop systems	• ______ • ______ • ______ • ______	• Relies on a human operator at all times. • Cannot be used in dangerous areas. • Processes can be affected by slow operator reactions.
Closed-loop systems	• Can be used in dangerous areas. • Quick and efficient. • Does not require a human operator to be present.	• ______ ______ • ______ ______ • ______

Unit 5, Lesson 7, Exercise B
Page 103

Look at the key and complete table 2 on page 103 with abbreviations and full words. Ask your partner to help you complete table 1.

pressure	P	speed	S
furnace	F	recorder	R
level	L	indicator	I
temperature	T	motor	M

Unit 5, Review, Exercise C
Page 110

Look at the diagram. Ask your partner these questions.

1 What is the purpose of the sphere?
2 What happens if the sphere handling valve is opened?
3 How does the meter work?

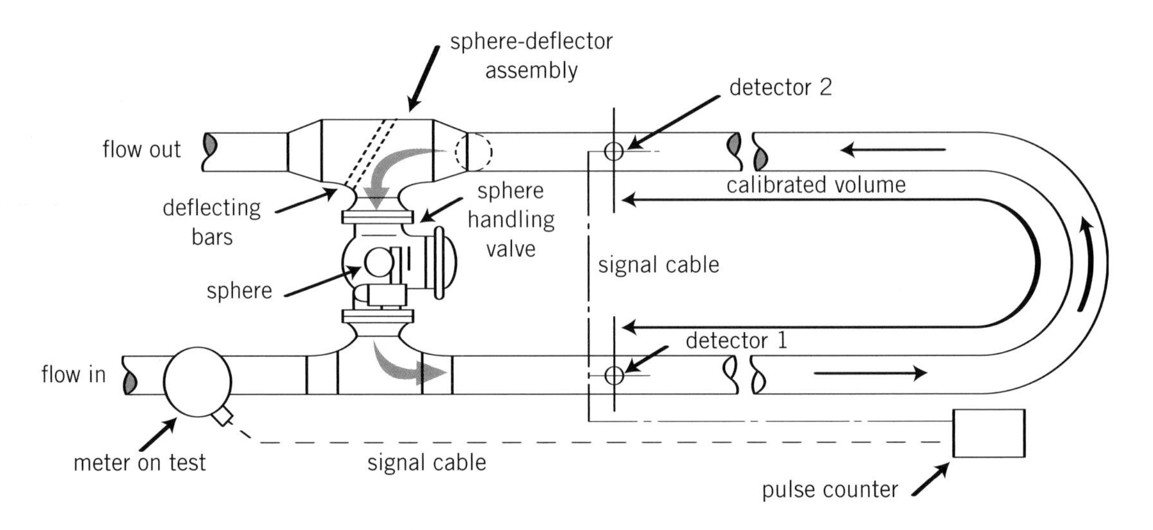

Unit 6, Lesson 6, Exercise B
Page 122

Complete the past tense table. Check your answers with your partner.

come	came	adjust	
control	controlled	be	
hear	heard	call	
initiate	initiated	increase	
reach	reached	shut down	
try	tried	stop	

Unit 6, Review, Exercise B
Page 132

Read the incident report.

In June 2003, at 2 p.m., an electrician was injured when he fell from a 16 ft aluminum tripod ladder. He was repairing the light fittings in a workshop which had an epoxy-coated concrete floor. The front single post of the ladder slid forward and the ladder collapsed with the electrician on it. He suffered severe head injuries as a result of the fall. He was taken to hospital and recovered consciousness 46 hours after the incident.

Unit 7, Review, Exercise D
Page 154

Read about different types of lubrication. Swap information about the advantages of each type, then discuss and write down some disadvantages.

Lubrication is vital – it reduces the friction between moving parts so they can work smoothly, which reduces wear. There are different grades of lubricant for different purposes.

Lubricant	Advantages	Disadvantages
Grease	• Suitable for high temperatures. • Seals out contaminants. • Can be used in difficult-to-access areas. • Easier to retain than oil.	• ______ • ______ • ______ • ______
Oil	• ______ • ______ • ______ • ______	• Often needs filters and coolers. • ______ • ______ • ______

Unit 8, Lesson 10, Exercise A
Page 174

Read the description of how a centrifuge works and label as much of the diagram below as you can. Then swap information with your partner and finish labelling the diagram.

A centrifuge works by spinning material at very high speed. This spinning pushes the heavier solid material to the outside of the vessel, separating it from the lighter particles. As the liquid passes through the coil in the centrifuge, it reaches the *parallel tail section*. Here, the heavier water particles move rapidly to the outer wall and leave the centrifuge via the *clean water outlet*. The lighter oil particles are dragged inwards and flow back up the middle of the centrifuge and flow out as the *oil reject stream* in the tapering section of the mechanism.

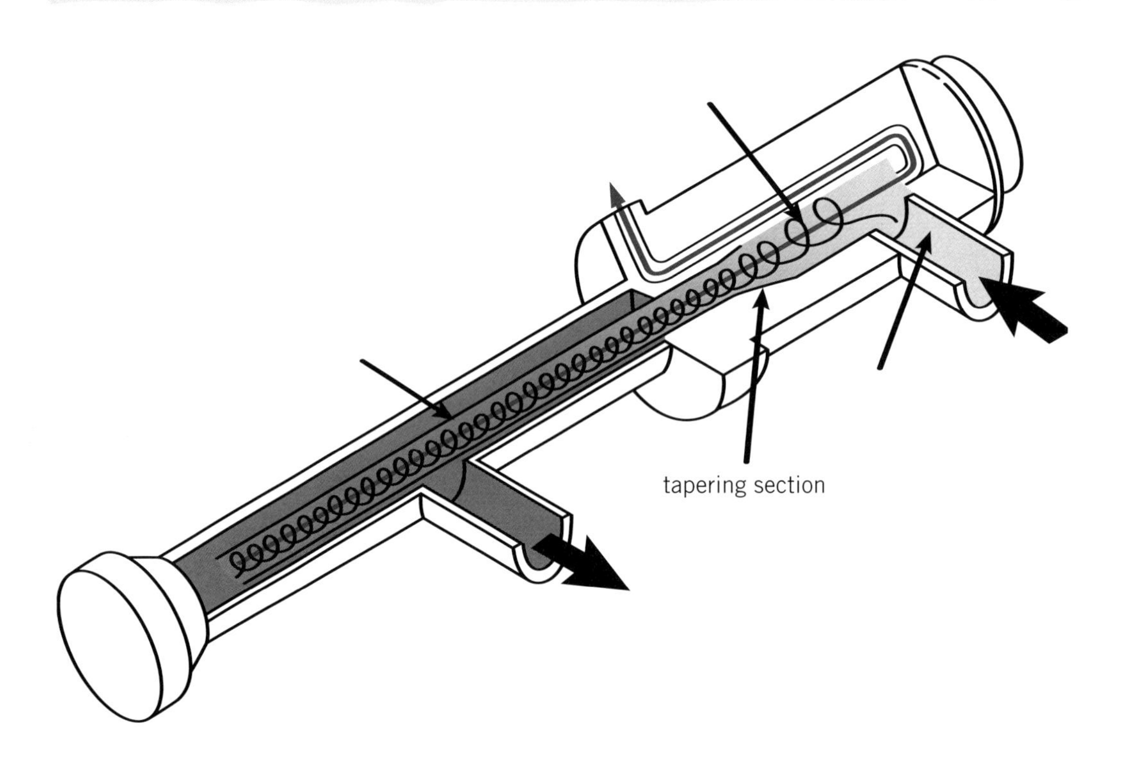

Review pairwork activities
Student 2

Review, Unit 1, Exercise C
Page 201

When your partner says a time, read out the corresponding letter.

 P	 T	 U	 E	 L	 O

 A	 N	 I	 S	 M	 R

Review, Unit 2, Exercise C
Page 202

Ask your partner for the missing information to complete the table.

No.	Symbol	Name	Atomic mass
1	H	Hydrogen	___
2	He	___	4.002602
3	___	Lithium	6.941
4	Be	Beryllium	___
5	B	___	10.811
6	___	Carbon	12.0107
7	N	Nitrogen	___
8	O	___	15.9994
9	___	Fluorine	18.9984032
10	___	___	20.1797
11	___	Sodium	22.989770
12	___	Magnesium	___
13	Al	___	26.981538
14	Si	___	___
15	___	Phosphorus	30.973761
16	___	___	___
17	___	___	___
18	Ar	Argon	39.948
21	Sc	Candium	4.955910
22	Ti	___	47.867
23	V	Vanadium	___
24	___	Chromium	51.9961
25	___	___	54.938049
26	Fe	Iron	55.8457
27	___	___	___
28	Ni	Nickel	58.6934
29	Cu	Copper	63.546
30	Zn	Zinc	___

Review, Unit 5, Exercise B
Page 207

Make up definitions for the words in your list using relative clauses. Read your definitions to your partner and see if they can guess the words.

1 an incident report
2 an HSE officer
3 calipers
4 the floor of a rig

Review, Unit 5, Exercise E
Page 208

Read the description of the gas turbine. Ask your partner questions to find which words complete the gaps in your text.

1 First, air is drawn into the ______________. It gets hot due to the compression.

2 The compressed air is fed into the burner can in the combustion section. Here, it is mixed with ______________ and it is ignited.

3 The flames from the ignited fuel/air mixture go through tubes into other burner cans.

4 The burning gas becomes more pressurized and is directed onto the ______________.

5 The turbine disc rotates from the force of the gas. The compressor is on the same shaft as the turbine disc, so this rotates also. This ensures that ______________.

6 After passing through the turbine disc, the hot gas strikes the load turbine rotor disc. The load turbine rotates the load rotor shaft, which is connected to ______________.

7 The hot gas is no longer needed, so it leaves the engine through the exhaust.

Glossary

alarm	a device that warns of danger by making a sound or giving a signal, e.g., a fire alarm, a smoke alarm, a security alarm
asphalt	a brownish-black solid or semi-solid which is a by-product of petroleum, used in paving, roofing and waterproofing
ATV	an all-terrain vehicle such as a quad bike
barge	a long boat for carrying freight that is usually pushed or pulled by another vessel
barge-type rig	a flat-bottomed rig, suitable for shallow waters, with the derrick over a moon pool in the centre of the barge
barrel	1 a large, wooden or metal container 2 a measurement used for crude oil: 1 barrel (bbl) equals 159 litres, 35 imperial gallons or 42 US gallons
bellows	an instrument that expands and contracts, drawing in air through a valve and pushing it out again
bit	the cutting part of the drilling equipment, made of a strong material such as diamonds
blowout	when the upward pressure in the well is so high that it overcomes the pressure of the drilling fluid and causes gas, oil or water to rise to the surface and escape in an uncontrolled manner
blowout preventer (BOP)	hydraulically-operated wellhead device designed to stop blowouts occurring
BOP stack	an assembly of blowout preventers and associated equipment mounted on the wellhead for the purpose of controlling pressure down the drill hole
bourdon tube	an instrument for measuring the pressure of steam or other gases
bubble cap	a metal cap covering a hole in a plate in a distillation tower; it lets rising vapour pass through the cap and condense on the plate

butterfly valve — a type of quick-opening valve which is opened and closed by a disc that pivots on a shaft inside the valve

cable — a rope made of wire, hemp or other strong fibres

capillary tube — a narrow tube that draws liquid through it against the force of gravity

casing — a steel pipe that lines the inside of a well to prevent unwanted fluids from entering the well and stops the walls from collapsing

centrifuge — an apparatus that spins material about a central axis in order to separate it into different constituents

check valve — a valve that permits flow in one direction only

circulating — the continuous process of pumping drilling mud through the drill string system during drilling operations

combustion engine — an engine, such as a gasoline piston engine or a diesel engine, where the fuel is burnt inside

compound — a substance where two or more elements are joined together and can only be separated by chemical means

compressor — equipment used to compress gas for reinjection into a well or for pumping through a pipeline

condenser — equipment that changes a material from a gas (vapour) to a liquid by cooling it

contamination — the act of soiling or polluting, e.g., by spilling substances (intentionally or accidentally) such as chemicals or biological organisms

crane — a machine for lifting and moving heavy objects, usually with a projecting arm or a jib and a rotating base; can be part of the derrick

crew — a group of people who work together, e.g., the seismic crew, the drilling crew

crude oil — unrefined petroleum; a thick, black mixture of hundreds of different compounds

cuttings	the fragments of rock cut up by the bit and brought to the surface in the drilling mud
debris	broken rock fragments and rubbish that accumulates, e.g., inside a pipeline
demulsifier	a chemical, mechanical or electrical system that breaks down emulsions
derrick	a tall triangular framework that is erected over an oil well to support equipment or raise and lower pipes
derrick monkey	a member of the drill crew who works at the top of the rig (also known as the derrick man)
diesel	a light oil that is used in diesel and other compression-ignition engines
discharge	something that releases or empties a substance, e.g., a discharge tank, discharge valve or discharge ball check in a pump
distillation	the process by which liquids and gases are separated or purified by heating and cooling them so that they vaporize, then condense, i.e., gaseous fuels, kerosene and gas oils
diver	someone who works underwater in a wet suit or diving apparatus
doodlebugger	someone who works on the seismic crew
downtime	the length of time that an operation is delayed, usually due to bad weather or mechanical failure
drain	a device that allows liquid to leave a container such as a tank
drill collar	the top of a well, usually a cemented section
driller	an experienced operator who heads the drilling crew and operates the drilling machinery

drilling mud	a mixture of minerals and chemicals in solution that is circulated through a well to control the pressure of the reservoir fluid, clean the drill system and carry cuttings out of the well (also known as drilling fluid)
drill stem	all parts of the assembly used for rotary drilling from the swivel to the bits, including the kelly, drill pipe and tool joints and drill collars
drill string	the column of heavy drill pipe lengths (each approximately 30 feet long) that are screwed together to form a tube connecting the drill bit to the drilling rig; the drill string rotates to drill the hole and carries the drilling fluid to the bit (also known as the drill pipe)
element	a substance which cannot be broken down into a simpler form, e.g., oxygen
emulsion	a combination of two liquids that do not mix, e.g., oil and water
fertilizers	organic materials and chemicals that are put on or dug into the soil to increase and improve plant growth
filter	a device used for separating and removing solids or suspended particles from liquids
fishing	an attempt to recover tools or drilling equipment (fish) lost or caught in a well
fixed platform	a (permanent) rig that is suitable for deeper waters (usually 50–300 feet) with the drilling rig installed on an underwater jacket (steel structure)
flammable	something that is easily ignited and capable of burning rapidly; inflammable
flare	an open flame used to burn off unwanted gas
float	a buoyant object used to hold things up in water or measure the level of liquid in a tank
floating roof tank	a large, open container where oil is stored
floorman	a member of the drilling crew who works on the derrick floor

flow meter — a meter that measures the quantity of a gas or liquid flowing through a pipe

foam —
1 bubbles of air or gas formed in or on the surface of a liquid
2 a light, porous, semirigid or spongy material used for insulation or shock absorption, as in packaging

foreman — someone who is leader of a work crew

fractional distillation — the process where crude oil is split into liquids of different boiling ranges (fractions) by distillation; the basic process that takes place in an oil refinery

gasoline — a flammable liquid fuel refined from crude petroleum and used principally in internal-combustion engines (also known as petrol)

gauge — an instrument for measuring and/or indicating quantity or level

generator — a machine that changes mechanical energy into electrical energy

geophones — cables with detectors (microphones) that pick up sound waves from under the ground and the ocean

grease — a thick, non-liquid lubricant that is used to protect components such as sealed bearings

gusher — an oil well that blows out of control due to a sudden increase in pressure

heater treater — a machine that heats an oil-and-water emulsion to separate it into different streams

hoist — pulleys and rope or chain used to lift heavy objects

hose — a flexible tube for carrying liquids or gases under pressure

HSE — a unit for heating an oil-and-water emulsion and then removing the water and gas

hydrocarbons	organic compounds composed of carbon and hydrogen which are the main components of petroleum and oil-based products
industrial fuel oil	heavy distillates of petroleum that must be heated to high temperatures before vaporization occurs; used as fuel in power stations, industry and shipping
injection well	a well used to inject gas or water into the reservoir rock to maintain pressure in secondary recovery
jack-up rig	a mobile rig with supporting legs that can be raised, or 'jacked up', when it moves to another location
jet fuel	high-grade kerosene and other light distillates of petroleum that ignite at low temperature and do not produce smoke
jughustler	the member of the seismic crew who operates the geophones
junk	equipment that is lost down a well (if it cannot be retrieved economically, the well is junked, i.e., plugged and left)
kelly	a hexagonal or square pipe about 45 feet long, attached to the top of the drill string and turned by the rotary table
kerosene	thin oil distilled from petroleum, used as a fuel for jet engines and for heating and cooking
kick	a sudden rush of high pressure into a well that sometimes happens while drilling
leak	the unwanted escape of liquid or gas from a container that has a hole or crack or is not fully sealed
lockouts	the procedure used when a piece of equipment is power-isolated in order for tags to be examined or repaired
loop systems	types of heating system that may be open or closed
LPG	liquefied petroleum gas is refined from crude oil and is mainly a mixture of propane and butane; it is gaseous at normal temperatures, but is made into a liquid by refrigeration or pressure to enable easy storage or transport

lubricant	a liquid or grease (often made from heavy liquid hydrocarbons) used to reduce friction in an engine or machine
metallurgist	a scientist who studies the properties of metals and ores to find oil deposits
meter	a device that measures and records amount or volume, e.g., the flow of gas, oil or electricity (see *flow meter*, *Venturi meter*, *turbine meter*)
mixture	a combination of two or more compounds that are not joined chemically and can be separated by physical means
molecule	the smallest particle of a compound that is capable of independent existence while retaining its individual properties
motorman	the man responsible for the care and operation of the drilling machinery
mud	see *drilling mud*
mud man	the man who maintains the mud systems on a drilling rig
multigrade lubricant	a lubricant that can operate at different temperatures without its viscosity being affected
nodding donkey	the nickname for the traditional type of pump used in onshore oil and natural gas fields; the up and down movement of the pump is similar to the movement of a donkey's head
offshore	applied to activities located or carried out at sea as opposed to on land
oilfield	a geographical area under which an oil reservoir lies
online	an oil well or piece of machinery or equipment is said to be online when it is in use or production as opposed to offline, when it is out of action
onshore	applied to activities located or carried out on land as opposed to at sea

on stream — a rig or oil well is on stream if it is in action and producing oil

operator — 1 a person or company that has authority to drill wells and undertake production if oil or gas is found
2 someone who uses a particular machine

overflow — an outlet that allows excess liquid to escape

paraffins — premium grade of kerosene burned in lamps and heaters; they can be in liquid form, or solid form as a wax

pentane — the saturated hydrocarbon (alkane) with five carbon atoms in its molecule; the fifth member of the paraffin series – a liquid under normal conditions

periodic table — a table that arranges the 116 elements into columns so that elements with similar properties are grouped together

permit — (usually written) authorization to work in a certain area, work with certain materials or machines, or work in a dangerous environment

petroleum — a generic name for hydrocarbons, including crude oil, natural gas liquids, natural gas and their products

P & ID — a process and instrument drawing diagram which shows complex industrial systems in a simple way

pig — a steel, plastic or foam device that is sent through a pipeline to clean it or check that there are no obstructions or breaks

pipeline — a pipe through which oil, oil products or gas is pumped between two points, either offshore or onshore

platform — an offshore structure from which oil wells are drilled

plug — 1 a plastic fitting with two or three metal prongs that connects an appliance to a power supply
2 a seal or other object that fills or covers a hole, e.g., a cement plug may be placed in an oil well to prevent high-pressure material escaping after it has been abandoned

pontoon	a hollow buoyancy tank used to support a semi-submersible rig, barge or other structure
PPE	personal protective equipment, including clothing and accessories that protect against hazards in the workplace, e.g., hard hats, safety goggles, boots, face masks
pressure gauge	a device that is used to measure pressure
pump	a mechanical device that transports liquid, air or gases from one vessel to another (the main types of pump are reciprocating, gear and centrifugal pumps)
recovery	the removal of oil from the Earth
refinery	a plant that separates the various components present in crude oil and converts them into products such as kerosene, diesel and petroleum
relay	an electronic device that gives a signal or activates a switch in response to a small change in voltage or current
reserves	a proportion of the oil and/or gas in a reservoir that can be removed using currently available techniques
reservoir	a place that stores a large amount of liquid; this could be a tank, a rock formation or rocks in an oilfield
rig	a large structure where machines are kept for drilling and producing oil and gas, and where oil workers live and work
rotameter	an instrument used for measuring the flow rate of a liquid or gas in a pipe
rotary table	a plate in the drill floor which is turned mechanically to provide the rotary action to the drill string which passes though its centre
rotor	a mechanical device that turns in a circular direction (rotates)
roughneck	a rig worker who handles the drill pipe and other equipment on the drill floor to assist the driller
round trip	the complete operation of pulling out the drill string from a well (for instance to change a bit) and then running it back into the well

roustabout	a general labourer on a rig who does routine cleaning and maintenance
sand trap	a device that removes smaller solid particles from the drilling mud
seal	a substance or device that is used to close a container or other object and prevent gas or liquid from escaping
seismic crew	workers who use acoustic methods of determining what is under the Earth or sea by sending shock waves into the various buried rock layers
semi-submersible	a kind of rig that is suitable for deep-water operations (usually 200–1,500 feet) and is supported by floating pontoons that are under the water
separation	the process of breaking up the fluid pumped out of an oil well into separate oil, water and gas streams
shale shaker	a device that removes rock cuttings from the drilling mud
solute	a substance that can be dissolved in liquid, usually water, e.g., salt
solvent	a substance, usually a liquid, that can dissolve other substances, e.g., cleaning fluids
spillage	an accidental release of oil or another liquid
spotter	the person who helps a crane operator use the crane safely by signalling what to do
strainer	a metal screen similar to a filter, but which separates larger, coarse particles from liquid
submersible drilling rig	an offshore drilling structure with part of the rig submerged and resting on the sea floor; commonly used only in shallow waters
tag number	a number used to indicate the function of, and identify, a particular piece of equipment
tank	a container that stores liquid for various purposes, e.g., a storage tank, supply tank or discharge tank

tanker	a ship or vehicle used to transport oil, refined products or liquefied gas
tension leg platform	a rig that is similar to a semi-submersible rig but is attached to the ocean floor by tensioned steel cables
terminal	an onshore installation that receives oil and/or gas from a pipeline or from tankers; not a refinery
thermostat	a device that controls temperature by responding to increased or decreased heat
trip the pipe	the action of removing the pipe from the hole and running it in again
turbine	a piece of equipment that has a shaft (rotor) that is steadily rotated by a stream of liquid or vapour as it comes into contact with a wheel
turbine meter	a meter that has a rotor at its centre; the movement of the rotor is measured to find the flow rate
ultrasonic	a device or machine is ultrasonic if it uses acoustic frequencies above the range that a human ear can detect, or above approximately 20,000 hertz
valve	a device that controls or regulates the flow of a liquid or gas in a pipe, e.g., a butterfly valve, gate valve, swing valve or check valve
vapour	a substance in a gaseous state
Venturi meter	an instrument for measuring flow rate by measuring the pressure in the pipeline at two different points
viscosity	a measure of how easily a liquid moves or flows; it usually decreases as the temperature rises
water injection	a process whereby treated water is pumped into the reservoir rock in order to raise or maintain the pressure and push crude oil towards the central production wells
wax	solid hydrocarbon which is present in some crude oils, especially in paraffin